Fifty Challenging Problems in Probability with Solutions

FREDERICK MOSTELLER
Professor of Mathematical Statistics
Harvard University

Dover Publications, Inc., New York

Published in Canada by General Publishing Company, Ltd., 30 Lesmill Road, Don Mills, Toronto, Ontario.
Published in the United Kingdom by Constable and Company, Ltd., 10 Orange Street, London WC2H 7EG.

This Dover edition, first published in 1987, in an unabridged and unaltered republication of the work first published by Addison-Wesley Publishing Company, Inc., Reading, MA, in 1965.

Manufactured in the United States of America
Dover Publications, Inc., 31 East 2nd Street, Mineola, N.Y. 11501

Library of Congress Cataloging-in-Publication Data

Mosteller, Frederick, 1916–
Fifty challenging problems in probability with solutions.

Reprint. Originally published: Reading, MA : Addison-Wesley, 1965. Originally published in series: A-W series in introductory college mathematics.
1. Probabilities—Problems, exercises, etc. I. Title.
QA273.25.M67 1987 519.2′076 86-32957
ISBN 0-486-65355-2 (pbk.)

Preface

This book contains 56 problems although only 50 are promised. A couple of the problems prepare for later ones; since tastes differ, some others may not challenge you; finally, six are discussed rather than solved. If you feel your capacity for challenge has not been exhausted, try proving the final remark in the solution of Problem 48. One of these problems has enlivened great parts of the research lives of many excellent mathematicians. Will you be the one who completes it? Probably not, but it hasn't been proved impossible.

Much of what I have learned, as well as much of my intellectual enjoyment, has come through problem solving. Through the years, I've found it more and more difficult to tell when I was working and when playing, for it has so often turned out that what I have learned playing with problems has been useful in my serious work.

In a problem, the great thing is the challenge. A problem can be challenging for many reasons: because the subject matter is intriguing, because the answer defies unsophisticated intuition, because it illustrates an important principle, because of its vast generality, because of its difficulty, because of a clever solution, or even because of the simplicity or beauty of the answer.

In this book, many of the problems are easy, some are hard. A very few require calculus, but a person without this equipment may enjoy the problem and its answer just the same. I have been more concerned about the challenge than about precisely limiting the mathematical level. In a few instances, where a special formula is needed that the reader may not have at his finger tips, or even in his repertoire, I have supplied it without ado. Stirling's approximation for the factorials (see Problem 18) and Euler's approximation for the partial sum of a harmonic series (see Problem 14) are two that stand out.

Perhaps the reader will be as surprised as I was to find that the numbers π, which relates diameters of circles to their circumferences, and e, which is the base of the natural logarithms, arise so often in probability problems.

In the Solutions section, the upper right corner of the odd-numbered pages carries the number of the last problem being discussed on that page. We hope that this may be of help in turning back and forth in the book. The pages are numbered at the bottom.

Anyone writing on problems in probability owes a great debt to the mathematical profession as a whole and probably to W. A. Whitworth and his book *Choice and chance* (Hafner Publishing Co., 1959, Reprint of fifth edition much enlarged, issued in 1901) in particular.

One of the pleasures of a preface is the opportunity it gives the author to acknowledge his debts to friends. To Robert E. K. Rourke goes the credit or blame for persuading me to assemble this booklet; and in many problems the wording has been improved by his advice. My old friends and critics Andrew Gleason, L. J. Savage, and John D. Williams helped lengthen the text by proposing additional problems for inclusion, by suggesting enticing extensions for some of the solutions, and occasionally by offering to exchange right for wrong; fortunately, I was able to resist only a few of these suggestions. In addition, I owe direct personal debts for suggestions, aid, and conversations to Kai Lai Chung, W. G. Cochran, Arthur P. Dempster, Bernard Friedman, John Garraty, John P. Gilbert, Leo Goodman, Theodore Harris, Olaf Helmer, J. L. Hodges, Jr., John G. Kemeny, Thomas Lehrer, Jess I. Marcum, Howard Raiffa, Herbert Scarf, George B. Thomas, Jr., John W. Tukey, Lester E. Dubins, and Cleo Youtz.

Readers who wish a systematic elementary development of probability may find helpful material in F. Mosteller, R. E. K. Rourke, G. B. Thomas, Jr., *Probability with statistical applications*, Addison-Wesley, Reading, Mass., 1961. In referring to this book in the text I have used the abbreviation PWSA. A shorter version is entitled *Probability and statistics*, and a still shorter one, *Probability: A first course*.

More advanced material can be found in the following: W. Feller, *An introduction to probability theory and its applications*, Wiley, New York; E. Parzen, *Modern probability theory and its applications*, Wiley, New York.

FREDERICK MOSTELLER

West Falmouth, Massachusetts
August, 1964

Contents

Fifty Challenging Problems in Probability

1. The Sock Drawer

A drawer contains red socks and black socks. When two socks are drawn at random, the probability that both are red is $\frac{1}{2}$. (a) How small can the number of socks in the drawer be? (b) How small if the number of black socks is even?

2. Successive Wins

To encourage Elmer's promising tennis career, his father offers him a prize if he wins (at least) two tennis sets in a row in a three-set series to be played with his father and the club champion alternately: father-champion-father or champion-father-champion, according to Elmer's choice. The champion is a better player than Elmer's father. Which series should Elmer choose?

3. The Flippant Juror

A three-man jury has two members each of whom independently has probability p of making the correct decision and a third member who flips a coin for each decision (majority rules). A one-man jury has probability p of making the correct decision. Which jury has the better probability of making the correct decision?

4. Trials until First Success

On the average, how many times must a die be thrown until one gets a 6?

5. Coin in Square

In a common carnival game a player tosses a penny from a distance of about 5 feet onto the surface of a table ruled in 1-inch squares. If the penny ($\frac{3}{4}$ inch in diameter) falls entirely inside a square, the player receives 5 cents but does not get his penny back; otherwise he loses his penny. If the penny lands on the table, what is his chance to win?

6. Chuck-a-Luck

Chuck-a-Luck is a gambling game often played at carnivals and gambling houses. A player may bet on any one of the numbers 1, 2, 3, 4, 5, 6. Three dice are rolled. If the player's number appears on one, two, or three of the dice, he receives respectively one, two, or three times his original stake plus his own money back; otherwise he loses his stake. What is the player's expected loss per unit stake? (Actually the player may distribute stakes on several numbers, but each such stake can be regarded as a separate bet.)

7. Curing the Compulsive Gambler

Mr. Brown always bets a dollar on the number 13 at roulette against the advice of Kind Friend. To help cure Mr. Brown of playing roulette, Kind Friend always bets Brown $20 at even money that Brown will be behind at the end of 36 plays. How is the cure working?

(Most American roulette wheels have 38 equally likely numbers. If the player's number comes up, he is paid 35 times his stake and gets his original stake back; otherwise he loses his stake.)

8. Perfect Bridge Hand

We often read of someone who has been dealt 13 spades at bridge. With a well-shuffled pack of cards, what is the chance that you are dealt a perfect hand (13 of one suit)? (Bridge is played with an ordinary pack of 52 cards, 13 in each of 4 suits, and each of 4 players is dealt 13.)

9. Craps

The game of craps, played with two dice, is one of America's fastest and most popular gambling games. Calculating the odds associated with it is an instructive exercise.

The rules are these. Only totals for the two dice count. The player throws the dice and wins at once if the total for the first throw is 7 or 11, loses at once if it is 2, 3, or 12. Any other throw is called his "point."* If the first throw is a point, the player throws the dice repeatedly until he either wins by throwing his point again or loses by throwing 7. What is the player's chance to win?

10. An Experiment in Personal Taste for Money

(a) An urn contains 10 black balls and 10 white balls, identical except for color. You choose "black" or "white." One ball is drawn at random, and if its color matches your choice, you get $10, otherwise nothing. Write down the maximum amount you are willing to pay to play the game. The game will be played just once.

(b) A friend of yours has available many black and many white balls, and he puts black and white balls into the urn to suit himself. You choose "black" or "white." A ball is drawn randomly from this urn. Write down the maximum amount you are willing to pay to play this game. The game will be played just once.

Problems without Structure (11 and 12)

Olaf Helmer and John Williams of The RAND Corporation have called my attention to a class of problems that they call "problems without structure," which nevertheless seem to have probabilistic features, though not in the usual sense.

11. Silent Cooperation

Two strangers are separately asked to choose one of the positive whole numbers and advised that if they both choose the same number, they both get a prize. If you were one of these people, what number would you choose?

---

*The throws have catchy names: for example, a total of 2 is Snake eyes, of 8, Eighter from Decatur, of 12, Boxcars. When an even point is made by throwing a pair, it is made "the hard way."

12. Quo Vadis?

Two strangers who have a private recognition signal agree to meet on a certain Thursday at 12 noon in New York City, a town familiar to neither, to discuss an important business deal, but later they discover that they have not chosen a meeting place, and neither can reach the other because both have embarked on trips. If they try nevertheless to meet, where should they go?

13. The Prisoner's Dilemma

Three prisoners, A, B, and C, with apparently equally good records have applied for parole. The parole board has decided to release two of the three, and the prisoners know this but not which two. A warder friend of prisoner A knows who are to be released. Prisoner A realizes that it would be unethical to ask the warder if he, A, is to be released, but thinks of asking for the name of *one* prisoner *other than himself* who is to be released. He thinks that before he asks, his chances of release are $\frac{2}{3}$. He thinks that if the warder says "B will be released," his own chances have now gone down to $\frac{1}{2}$, because either A and B or B and C are to be released. And so A decides not to reduce his chances by asking. However, A is mistaken in his calculations. Explain.

14. Collecting Coupons

Coupons in cereal boxes are numbered 1 to 5, and a set of one of each is required for a prize. With one coupon per box, how many boxes on the average are required to make a complete set?

15. The Theater Row

Eight eligible bachelors and seven beautiful models happen randomly to have purchased single seats in the same 15-seat row of a theater. On the average, how many pairs of adjacent seats are ticketed for marriageable couples?

16. Will Second-Best Be Runner-Up?

A tennis tournament has 8 players. The number a player draws from a hat decides his first-round rung in the tournament ladder. See diagram.

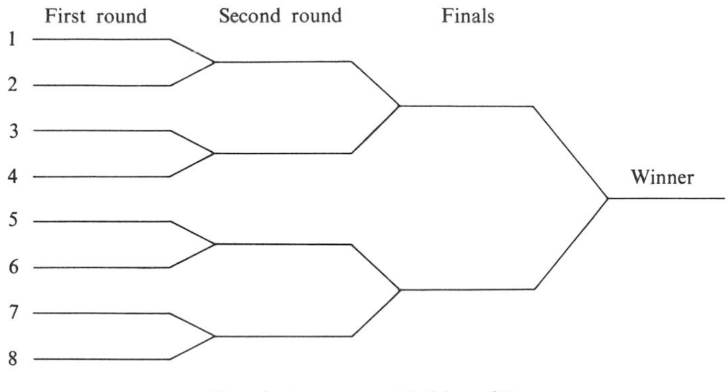

Tennis tournament ladder of 8.

Suppose that the best player always defeats the next best and that the latter always defeats all the rest. The loser of the finals gets the runner-up cup. What is the chance that the second-best player wins the runner-up cup?

17. Twin Knights

(a) Suppose King Arthur holds a jousting tournament where the jousts are in pairs as in a tennis tournament. See Problem 16 for tournament ladder. The 8 knights in the tournament are evenly matched, and they include the twin knights Balin and Balan.* What is the chance that the twins meet in a match during the tournament?

(b) Replace 8 by 2^n in the above problem. Now what is the chance that they meet?

18. An Even Split at Coin Tossing

When 100 coins are tossed, what is the probability that exactly 50 are heads?

*According to Arthurian legend, they were so evenly matched that on another occasion they slew each other.

19. Isaac Newton Helps Samuel Pepys

Pepys wrote Newton to ask which of three events is more likely: that a person get (a) at least 1 six when 6 dice are rolled, (b) at least 2 sixes when 12 dice are rolled, or (c) at least 3 sixes when 18 dice are rolled. What is the answer?

20. The Three-Cornered Duel

A, *B*, and *C* are to fight a three-cornered pistol duel. All know that *A*'s chance of hitting his target is 0.3, *C*'s is 0.5, and *B* never misses. They are to fire at their choice of target in succession in the order *A*, *B*, *C*, cyclically (but a hit man loses further turns and is no longer shot at) until only one man is left unhit. What should *A*'s strategy be?

21. Should You Sample with or without Replacement?

Two urns contain red and black balls, all alike except for color. Urn *A* has 2 reds and 1 black, and Urn *B* has 101 reds and 100 blacks. An urn is chosen at random, and you win a prize if you correctly name the urn on the basis of the evidence of two balls drawn from it. After the first ball is drawn and its color reported, you can decide whether or not the ball shall be replaced before the second drawing. How do you order the second drawing, and how do you decide on the urn?

22. The Ballot Box

In an election, two candidates, Albert and Benjamin, have in a ballot box a and b votes respectively, $a > b$, for example, 3 and 2. If ballots are randomly drawn and tallied, what is the chance that at least once after the first tally the candidates have the same number of tallies?

23. Ties in Matching Pennies

Players *A* and *B* match pennies N times. They keep a tally of their gains and losses. After the first toss, what is the chance that at no time during the game will they be even?

24. The Unfair Subway

Marvin gets off work at random times between 3 and 5 P.M. His mother lives uptown, his girl friend downtown. He takes the first subway that comes in either direction and eats dinner with the one he is first delivered to. His mother complains that he never comes to see her, but he says she has a 50-50 chance. He has had dinner with her twice in the last 20 working days. Explain.

25. Lengths of Random Chords

If a chord is selected at random on a fixed circle, what is the probability that its length exceeds the radius of the circle?

26. The Hurried Duelers

Duels in the town of Discretion are rarely fatal. There, each contestant comes at a random moment between 5 A.M. and 6 A.M. on the appointed day and leaves exactly 5 minutes later, honor served, unless his opponent arrives within the time interval and then they fight. What fraction of duels lead to violence?

27. Catching the Cautious Counterfeiter

(a) The king's minter boxes his coins 100 to a box. In each box he puts 1 false coin. The king suspects the minter and from each of 100 boxes draws a random coin and has it tested. What is the chance the minter's peculations go undetected?

(b) What if both 100's are replaced by n?

28. Catching the Greedy Counterfeiter

The king's minter boxes his coins n to a box. Each box contains m false coins. The king suspects the minter and randomly draws 1 coin from each of n boxes and has these tested. What is the chance that the sample of n coins contains exactly r false ones?

29. Moldy Gelatin

Airborne spores produce tiny mold colonies on gelatin plates in a laboratory. The many plates average 3 colonies per plate. What fraction of plates has exactly 3 colonies? If the average is a large integer m, what fraction of plates has exactly m colonies?

30. Evening the Sales

A bread salesman sells on the average 20 cakes on a round of his route. What is the chance that he sells an even number of cakes? (We assume the sales follow the Poisson distribution.)

Birthday Problems (31, 32, 33, 34)

31. Birthday Pairings

What is the least number of persons required if the probability exceeds $\frac{1}{2}$ that two or more of them have the same birthday? (Year of birth need not match.)

32. Finding Your Birthmate

You want to find someone whose birthday matches yours. What is the least number of strangers whose birthdays you need to ask about to have a 50-50 chance?

33. Relating the Birthday Pairings and Birthmate Problems

If r persons compare birthdays in the pairing problem, the probability is P_R that at least 2 have the same birthday. What should n be in the personal birthmate problem to make your probability of success approximately P_R?

34. Birthday Holidays

Labor laws in Erewhon require factory owners to give every worker a holiday whenever one of them has a birthday and to hire without discrimination on grounds of birthdays. Except for these holidays they work a 365-day year. The owners want to maximize the expected total number of man-days worked per year in a factory. How many workers do factories have in Erewhon?

35. The Cliff-Hanger

From where he stands, one step toward the cliff would send the drunken man over the edge. He takes random steps, either toward or away from the cliff. At any step his probability of taking a step away is $\frac{2}{3}$, of a step toward the cliff $\frac{1}{3}$. What is his chance of escaping the cliff?

36. Gambler's Ruin

Player M has \$1, and Player N has \$2. Each play gives one of the players \$1 from the other. Player M is enough better than Player N that he wins $\frac{2}{3}$ of the plays. They play until one is bankrupt. What is the chance that Player M wins?

37. Bold Play vs. Cautious Play

At Las Vegas, a man with \$20 needs \$40, but he is too embarrassed to wire his wife for more money. He decides to invest in roulette (which he doesn't enjoy playing) and is considering two strategies: bet the \$20 on "evens" all at once and quit if he wins or loses, or bet on "evens" one dollar at a time until he has won or lost \$20. Compare the merits of the strategies.

38. The Thick Coin

How thick should a coin be to have a $\frac{1}{3}$ chance of landing on edge?

The next few problems depend on the *Principle of Symmetry*. See pages 59–60.

39. The Clumsy Chemist

In a laboratory, each of a handful of thin 9-inch glass rods had one tip marked with a blue dot and the other with a red. When the laboratory assistant tripped and dropped them onto the concrete floor, many broke into three pieces. For these, what was the average length of the fragment with the blue dot?

40. The First Ace

Shuffle an ordinary deck of 52 playing cards containing four aces. Then turn up cards from the top until the first ace appears. On the average, how many cards are required to produce the first ace?

41. The Locomotive Problem

(a) A railroad numbers its locomotives in order, 1, 2, . . . , N. One day you see a locomotive and its number is 60. Guess how many locomotives the company has.

(b) You have looked at 5 locomotives and the largest number observed is 60. Again guess how many locomotives the company has.

42. The Little End of the Stick

(a) If a stick is broken in two at random, what is the average length of the smaller piece?

(b) (For calculus students.) What is the average ratio of the smaller length to the larger?

43. The Broken Bar

A bar is broken at random in two places. Find the average size of the smallest, of the middle-sized, and of the largest pieces.

44. Winning an Unfair Game

A game consists of a sequence of plays; on each play either you or your opponent scores a point, you with probability p (less than $\frac{1}{2}$), he with probability $1 - p$. The number of plays is to be even—2 or 4 or 6 and so on. To win the game you must get *more than* half the points. You know p, say 0.45, and you get a prize if you win. You get to choose in advance the number of plays. How many do you choose?

Matching Problems (45 and 46)

45. Average Number of Matches

The following are two versions of the matching problem:

(a) From a shuffled deck, cards are laid out on a table one at a time, face up from left to right, and then another deck is laid out so that each of its cards is beneath a card of the first deck. What is the average number of matches of the card above and the card below in repetitions of this experiment?

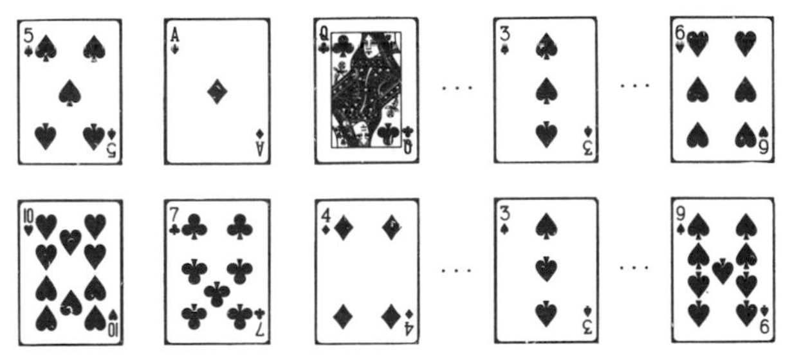

(b) A typist types letters and envelopes to n different persons. The letters are randomly put into the envelopes. On the average, how many letters are put into their own envelopes?

46. Probabilities of Matches

Under the conditions of the previous matching problem, what is the probability of exactly r matches?

47. Choosing the Largest Dowry

The king, to test a candidate for the position of wise man, offers him a chance to marry the young lady in the court with the largest dowry. The amounts of the dowries are written on slips of paper and mixed. A slip is drawn at random and the wise man must decide whether that is the largest dowry or not. If he decides it is, he gets the lady and her dowry if he is correct; otherwise he gets nothing. If he decides against the amount written on the first slip, he must choose or refuse the next slip, and so on until he chooses one or else the slips are exhausted. In all, 100 attractive young ladies participate, each with a different dowry. How should the wise man make his decision?

In the previous problem the wise man has no information about the distribution of the numbers. In the next he knows exactly.

48. Choosing the Largest Random Number

As a second task, the king wants the wise man to choose the largest number from among 100, by the same rules, but this time the numbers on the slips are randomly drawn from the numbers from 0 to 1 (random numbers, uniformly distributed). Now what should the wise man's strategy be?

49. Doubling Your Accuracy

An unbiased instrument for measuring distances makes random errors whose distribution has standard deviation σ. You are allowed two measurements all told to estimate the lengths of two cylindrical rods, one clearly longer than the other. Can you do better than to take one measurement on each rod? (An unbiased instrument is one that on the average gives the true measure.)

50. Random Quadratic Equations

What is the probability that the quadratic equation

$$x^2 + 2bx + c = 0$$

has real roots?

51. Two-Dimensional Random Walk

Starting from an origin O, a particle has a 50-50 chance of moving 1 step north or 1 step south, and also a 50-50 chance of moving 1 step east or 1 step west. After the step is taken, the move is repeated from the new position and so on indefinitely. What is the chance that the particle returns to the origin?

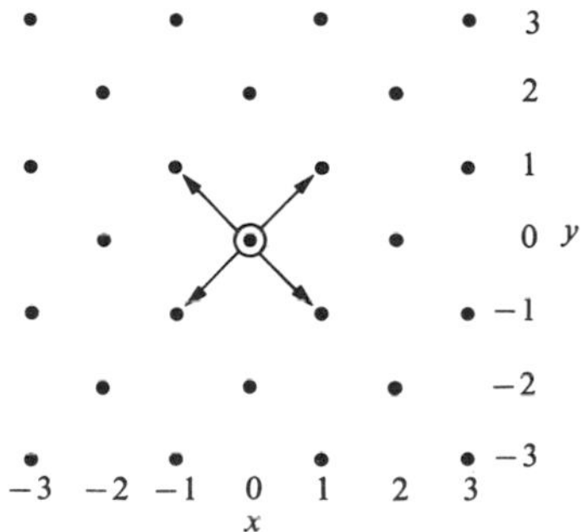

Part of lattice of points traveled by particles in the two-dimensional random walk problem. At each move the particle goes one step northeast, northwest, southeast, or southwest from its current position, the directions being equally likely.

52. Three-Dimensional Random Walk

As in the two-dimensional walk, a particle starts at an origin O in three-space. Think of the origin as centered in a cube 2 units on a side. One move in this walk sends the particle with equal likelihood to one of the *eight corners* of the cube. Thus, at every move the particle has a 50-50 chance of moving one unit up or down, one unit east or west, and one unit north or south. If the walk continues forever, find the fraction of particles that return to the origin.

53. Buffon's Needle

A table of infinite expanse has inscribed on it a set of parallel lines spaced $2a$ units apart. A needle of length $2l$ (smaller than $2a$) is twirled and tossed on the table. What is the probability that when it comes to rest it crosses a line?

54. Buffon's Needle with Horizontal and Vertical Rulings

Suppose we toss a needle of length $2l$ (less than 1) on a grid with both horizontal and vertical rulings spaced one unit apart. What is the mean number of lines the needle crosses? (I have dropped the $2a$ for the spacing because we might as well think of the length of the needle as measured in units of spacing.)

55. Long Needles

In the previous problem let the needle be of arbitrary length, then what is the mean number of crosses?

56. Molina's Urns

Two urns contain the same total numbers of balls, some blacks and some whites in each. From each urn are drawn n (≥ 3) balls with replacement. Find the number of drawings and the composition of the two urns so that the probability that all white balls are drawn from the first urn is equal to the probability that the drawing from the second is either all whites or all blacks.

Solutions for Fifty Challenging Problems in Probability

1. The Sock Drawer

A drawer contains red socks and black socks. When two socks are drawn at random, the probability that both are red is $\frac{1}{2}$. (a) How small can the number of socks in the drawer be? (b) How small if the number of black socks is even?

Solution for The Sock Drawer

Just to set the pattern, let us do a numerical example first. Suppose there were 5 red and 2 black socks; then the probability of the first sock's being red would be $5/(5 + 2)$. If the first were red, the probability of the second's being red would be $4/(4 + 2)$, because one red sock has already been removed. The product of these two numbers is the probability that both socks are red:

$$\frac{5}{5+2} \times \frac{4}{4+2} = \frac{5(4)}{7(6)} = \frac{10}{21}.$$

This result is close to $\frac{1}{2}$, but we need exactly $\frac{1}{2}$. Now let us go at the problem algebraically.

Let there be r red and b black socks. The probability of the first sock's being red is $r/(r + b)$; and if the first sock is red, the probability of the second's being red now that a red has been removed is $(r - 1)/(r + b - 1)$. Then we require the probability that both are red to be $\frac{1}{2}$, or

$$\frac{r}{r+b} \times \frac{r-1}{r+b-1} = \frac{1}{2}.$$

One could just start with $b = 1$ and try successive values of r, then go to $b = 2$ and try again, and so on. That would get the answers quickly. Or we could play along with a little more mathematics. Notice that

$$\frac{r}{r+b} > \frac{r-1}{r+b-1}, \qquad \text{for } b > 0.$$

Therefore we can create the inequalities

$$\left(\frac{r}{r+b}\right)^2 > \frac{1}{2} > \left(\frac{r-1}{r+b-1}\right)^2.$$

Taking square roots, we have, for $r > 1$,

$$\frac{r}{r+b} > \frac{1}{\sqrt{2}} > \frac{r-1}{r+b-1}.$$

From the first inequality we get

$$r > \frac{1}{\sqrt{2}}(r+b)$$

or

$$r > \frac{1}{\sqrt{2}-1}b = (\sqrt{2}+1)b.$$

From the second we get

$$(\sqrt{2}+1)b > r-1$$

or all told

$$(\sqrt{2}+1)b+1 > r > (\sqrt{2}+1)b.$$

For $b = 1$, r must be greater than 2.414 and less than 3.414, and so the candidate is $r = 3$. For $r = 3$, $b = 1$, we get

$$P(\text{2 red socks}) = \tfrac{3}{4} \cdot \tfrac{2}{3} = \tfrac{1}{2}.$$

And so the smallest number of socks is 4.

Beyond this we investigate even values of b.

b	r is between	eligible r	P(2 red socks)
2	5.8, 4.8	5	$\frac{5(4)}{7(6)} \neq \frac{1}{2}$
4	10.7, 9.7	10	$\frac{10(9)}{14(13)} \neq \frac{1}{2}$
6	15.5, 14.5	15	$\frac{15(14)}{21(20)} = \frac{1}{2}$

And so 21 socks is the smallest number when b is even. If we were to go on and ask for further values of r and b so that the probability of two red socks is $\frac{1}{2}$, we would be wise to appreciate that this is a problem in the theory of numbers. It happens to lead to a famous result in Diophantine Analysis obtained from Pell's equation.* Try $r = 85$, $b = 35$.

*See for example, W. J. LeVeque, *Elementary theory of numbers*, Addison-Wesley, Reading, Mass., 1962, p. 111.

2. Successive Wins

To encourage Elmer's promising tennis career, his father offers him a prize if he wins (at least) two tennis sets in a row in a three-set series to be played with his father and the club champion alternately: father-champion-father or champion-father-champion, according to Elmer's choice. The champion is a better player than Elmer's father. Which series should Elmer choose?

Solution for Successive Wins

Since the champion plays better than the father, it seems reasonable that fewer sets should be played with the champion. On the other hand, the middle set is the key one, because Elmer cannot have two wins in a row without winning the middle one. Let C stand for champion, F for father, and W and L for a win and a loss by Elmer. Let f be the probability of Elmer's winning any set from his father, c the corresponding probability of winning from the champion. The table shows the only possible prize-winning sequences together with their probabilities, given independence between sets, for the two choices.

	Father first				Champion first			
Set with:	F	C	F	Probability	C	F	C	Probability
	W	W	W	fcf	W	W	W	cfc
	W	W	L	$fc(1-f)$	W	W	L	$cf(1-c)$
	L	W	W	$(1-f)cf$	L	W	W	$(1-c)fc$
Totals				$fc(2-f)$				$fc(2-c)$

Since Elmer is more likely to best his father than to best the champion, f is larger than c, and $2 - f$ is smaller than $2 - c$, and so Elmer should choose CFC. For example, for $f = 0.8$, $c = 0.4$, the chance of winning the prize with FCF is 0.384, that for CFC is 0.512. Thus the importance of winning the middle game outweighs the disadvantage of playing the champion twice.

Many of us have a tendency to suppose that the higher the expected number of successes, the higher the probability of winning a prize, and often this supposition is useful. But occasionally a problem has special conditions that destroy this reasoning by analogy. In our problem the expected number of wins under CFC is $2c + f$, which is less than the expected number of wins for FCF, $2f + c$. In our example with $f = 0.8$ and $c = 0.4$, these means are 1.6 and 2.0 in that order. This opposition of answers gives the problem its flavor. The idea of independent events is explained in PWSA, pp. 81–84.

3. The Flippant Juror

A three-man jury has two members each of whom independently has probability p of making the correct decision and a third member who flips a coin for each decision (majority rules). A one-man jury has probability p of making the correct decision. Which jury has the better probability of making the correct decision?

Solution for The Flippant Juror

The two juries have the same chance of a correct decision. In the three-man jury, the two serious jurors agree on the correct decision in the fraction $p \times p = p^2$ of the cases, and for these cases the vote of the joker with the coin does not matter. In the other correct decisions by the three-man jury, the serious jurors vote oppositely, and the joker votes with the "correct" juror. The chance that the serious jurors split is $p(1 - p) + (1 - p)p$ or $2p(1 - p)$. Halve this because the coin favors the correct side half the time. Finally, the total probability of a correct decision by the three-man jury is $p^2 + p(1 - p) = p^2 + p - p^2 = p$, which is identical with the probability given for the one-man jury.

4. Trials until First Success

On the average, how many times must a die be thrown until one gets a 6?

Solutions for Trials until First Success

It seems obvious that it must be 6. To check, let p be the probability of a 6 on a given trial. Then the probabilities of success for the first time on each trial are (let $q = 1 - p$):

Trial	Probability of success on trial
1	p
2	pq
3	pq^2
.	.
.	.
.	.

The sum of the probabilities is

$$
\begin{aligned}
p + pq + pq^2 + \cdots &= p(1 + q + q^2 + \cdots) \\
&= p/(1 - q) = p/p = 1.
\end{aligned}
$$

The mean number of trials, m, is by definition,

$$m = p + 2pq + 3pq^2 + 4pq^3 + \cdots.$$

Note that our usual trick for summing a geometric series works:

$$qm = \quad pq + 2pq^2 + 3pq^3 + \cdots.$$

Subtracting the second expression from the first gives

$$m - qm = p + pq + pq^2 + \cdots,$$

or

$$m(1 - q) = 1.$$

Consequently,

$$mp = 1, \qquad \text{and} \qquad m = 1/p.$$

In our example, $p = \frac{1}{6}$, and so $m = 6$, as seemed obvious.

I wanted to do the above algebra in detail because we come up against geometric distributions repeatedly. But a beautiful way to do this problem is to notice that when the first toss is a failure, the mean number of tosses required is $1 + m$, and when the first toss is a success, the mean number is 1. Then $m = p(1) + q(1 + m)$, or $m = 1 + qm$, and

$$m = 1/p.$$

5. Coin in Square

In a common carnival game a player tosses a penny from a distance of about 5 feet onto the surface of a table ruled in 1-inch squares. If the penny ($\frac{3}{4}$ inch in diameter) falls entirely inside a square, the player receives 5 cents but does not get his penny back; otherwise he loses his penny. If the penny lands on the table, what is his chance to win?

Solution for Coin in Square

When we toss the coin onto the table, some positions for the center of the coin are more likely than others, but over a very small square we can regard the probability distribution as uniform. This means that the probability that the center falls into any region of a square is proportional to the area of the region, indeed, is the area

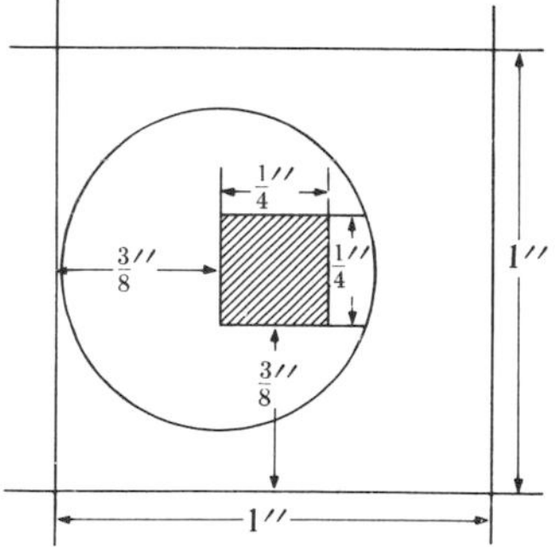

Shaded area shows where center of coin must fall for player to win.

of the region divided by the area of the square. Since the coin is $\frac{3}{8}$ inch in radius, its center must not land within $\frac{3}{8}$ inch of any edge if the player is to win. This restriction generates a square of side $\frac{1}{4}$ inch within which the center of the coin must lie for the coin to be in the square. Since the probabilities are proportional to areas, the probability of winning is $(\frac{1}{4})^2 = \frac{1}{16}$. Of course, since there is a chance that the coin falls off the table altogether, the total probability of winning is smaller still. Also the squares can be made smaller by merely thickening the lines. If the lines are $\frac{1}{16}$ inch wide, the winning central area reduces the probability to $(\frac{3}{16})^2 = \frac{9}{256}$ or less than $\frac{1}{28}$.

6. Chuck-a-Luck

Chuck-a-Luck is a gambling game often played at carnivals and gambling houses. A player may bet on any one of the numbers 1, 2, 3, 4, 5, 6. Three dice are rolled. If the player's number appears on one, two, or three of the dice, he receives respectively one, two, or three times his original stake plus his own money back; otherwise he loses his stake. What is the player's expected loss per unit stake? (Actually the player may distribute stakes on several numbers, but each such stake can be regarded as a separate bet.)

Solution for Chuck-a-Luck

Let us compute the losses incurred (a) when the numbers on the three dice are different, (b) when exactly two are alike, and (c) when all three are alike. An easy attack is to suppose that you place a unit stake on each of the six numbers, thus betting six units in all. Suppose the roll produces three different numbers, say 1, 2, 3. Then the house takes the three unit stakes on the losing numbers 4, 5, 6 and pays off the three winning numbers 1, 2, 3. The house won nothing, and you won nothing. That result would be the same for any roll of three *different* numbers.

Next suppose the roll of the dice results in two of one number and one of a second, say 1, 1, 2. Then the house can use the stakes on numbers 3 and 4 to pay off the stake on number 1, and the stake on number 5 to pay off that on number 2. This leaves the stake on number 6 for the house. The house won one unit, you lost one unit, or per unit stake you lost $\frac{1}{6}$.

Suppose the three dice roll the same number, for example, 1, 1, 1. Then the house can pay the triple odds from the stakes placed on 2, 3, 4 leaving those on 5 and 6 as house winnings. The loss per unit stake then is $\frac{2}{6}$. Note that when a roll produces a multiple payoff the players are losing the most on the average.

To find the expected loss per unit stake in the whole game, we need to weight the three kinds of outcomes by their probabilities. If we regard the three dice as distinguishable—say red, green, and blue—there are $6 \times 6 \times 6 = 216$ ways for them to fall.

In how many ways do we get three different numbers? If we take them in order, 6 possibilities for the red, then for each of these, 5 for the green since it must not match the red, and for each red-green pair, 4 ways for the blue since it must not match either of the others, we get $6 \times 5 \times 4 = 120$ ways.

For a moment skip the case where exactly two dice are alike and go on to three alike. There are just 6 ways because there are 6 ways for the red to fall and only 1 way for each of the others since they must match the red.

This means that there are $216 - 126 = 90$ ways for them to fall two alike and one different. Let us check that directly. There are three main patterns that give two alike: red-green alike, red-blue alike, or green-blue alike. Count the number of ways for one of these, say red-green alike, and then multiply by three. The red can be thrown 6 ways, then the green 1 way to match, and the blue 5 ways to fail to match, or 30 ways. All told then we have $3 \times 30 = 90$ ways, checking the result we got by subtraction.

We get the expected loss by weighting each loss by its probability and summing as follows:

none alike	2 alikc	3 alike

$$\frac{120}{216} \times 0 + \frac{90}{216} \times \frac{1}{6} + \frac{6}{216} \times \frac{2}{6} = \frac{17}{216} \approx 0.079.*$$

Thus you lose about 8% per play. Considering that a play might take half a minute and that government bonds pay you less than 4% interest for a year, the attrition can be regarded as fierce.

This calculation is for regular dice. Sometimes a spinning wheel with a pointer is used with sets of three numbers painted in segments around the edge of the wheel. The sets do not correspond perfectly to the frequencies given by the dice. In such wheels I have observed that the multiple payoffs are more frequent than for the dice, and therefore the expected loss to the bettor greater.

7. Curing the Compulsive Gambler

Mr. Brown always bets a dollar on the number 13 at roulette against the advice of Kind Friend. To help cure Mr. Brown of playing roulette, Kind Friend always bets Brown $20 at even money that Brown will be behind at the end of 36 plays. How is the cure working?

(Most American roulette wheels have 38 equally likely numbers. If the player's number comes up, he is paid 35 times his stake and gets his original stake back; otherwise he loses his stake.)

---

*The sign $\approx$ means "approximately equals."

Solution for Curing the Compulsive Gambler

If Mr. Brown wins once in 36 turns, he is even with the casino. His probability of losing all 36 times is $(\frac{37}{38})^{36} \approx 0.383$. In a single turn his expectation is

$$35(\tfrac{1}{38}) - 1(\tfrac{37}{38}) = -\tfrac{2}{38} \text{ (dollars)},$$

and in 36 turns

$$-\frac{2(36)}{38} \approx -1.89 \text{ (dollars)}.$$

Against Kind Friend, Mr. Brown has an expectation of

$$+20(0.617) - 20(0.383) \approx +4.68 \text{ (dollars)}.$$

And so all told Mr. Brown gains $+4.68 - 1.89 = +2.79$ dollars per 36 trials; he is finally making money at roulette. Possibly Kind Friend will be cured first. Of course, when Brown loses all 36, he is out $56, which may jolt him a bit.

8. Perfect Bridge Hand

We often read of someone who has been dealt 13 spades at bridge. With a well-shuffled pack of cards, what is the chance that you are dealt a perfect hand (13 of one suit)? (Bridge is played with an ordinary pack of 52 cards, 13 in each of 4 suits, and each of 4 players is dealt 13.)

Solution for Perfect Bridge Hand

The chances are mighty slim. Since the cards are well shuffled, we might as well deal your 13 off the top. To get 13 of one suit you can start with any card, and thereafter you are restricted to the same suit. So the number of ways to be dealt 13 of one suit is

$$52 \times 12 \times 11 \times 10 \times 9 \times 8 \times 7 \times 6 \times 5 \times 4 \times 3 \times 2 \times 1 = 52 \times 12!.$$

The total number of ways to get a bridge hand is

$$52 \times 51 \times 50 \times 49 \times 48 \times 47 \times 46 \times 45 \times 44 \times 43 \times 42 \times 41 \times 40 = 52!/39!.$$

The desired probability is $52 \times 12!/(52!/39!) = 12!39!/51!$. The reciprocal gives odds to 1 against. From 5-place tables of logarithms of factorials

(PWSA, p. 431) we have

$$\begin{array}{llll}
\log 12! & = \ \ 8.68034 & \log 51! & = 66.19065 \\
\log 39! & = \underline{46.30959} & \log(12!39!) & = \underline{54.98993} \\
\log(12!39!) & = 54.98993 & \log(12!39!/51!) & = 11.20072
\end{array}$$

$$\text{antilog: } 1.588 \times 10^{11}$$

In calculations of this kind, people sometimes get lost in the maze of exact figures. What matters here is that there is about one chance in 160 billion of a particular person's being dealt a perfect hand on a single deal. How often should we hear of it? Let's be generous and say that 10 million people play bridge in the United States of America and that each plays 10 hands a day every day of the year (equivalent to about two long sessions each week). That would give $36\frac{1}{2}$ billion hands a year, and so we expect about one perfect hand every 4 years, some of which would not be publicly reported. Even twice as many people playing twice as much would give only one such hand a year.

How does one account for the much higher frequency with which perfect hands are reported? Several things contribute. New decks have cards grouped by suits, and inadequate shuffling could account for some perfect hands. (A widely reported hand where all four players received perfect hands was the first hand dealt from a new deck.)

When we discuss very rare events, we have to worry about outrageous occurrences. No doubt quite a few reports owe their origin to pranks. Wouldn't grandma be surprised if she had 13 hearts for Valentine's Day? Let's arrange it, but we'll tell her later it was all a joke. Grandma takes her bridge seriously. When it turns out that grandma is overwhelmed, has called her relatives, bridge friends, and the reporters, news of a joke would be most unwelcome, and the easy course for the prankster is silence. Perhaps a few reports are made up out of whole cloth. It seems unlikely that this sort of hand would arise from accomplished cheating because it draws too much attention to the recipient and his partner.

N. T. Gridgeman discusses reports of perfect deals where all four players get 13 cards of one suit in "The mystery of the missing deal," *American Statistician*, Vol. 18, No. 1, Feb. 1964, pp. 15–16, and there is further correspondence in "Letters to the Editor," pp. 30–31, in the April, 1964 issue of that journal.

A slightly different way to compute this probability is to use binomial coefficients. They count the number of different ways to arrange a elements of one kind and b elements of another in a row. For example, 3 a's and 2 b's can be arranged in 10 ways, as the reader can verify on his fingers starting with *aaabb* and ending with *bbaaa*. The binomial coefficient is written $\binom{5}{2}$, meaning the number of ways to arrange 5 things, 2 of one kind, 3 of another. Its numerical value is given in terms of factorials:

$$\binom{5}{2} = \frac{5!}{2!3!} = \frac{5 \times 4 \times 3 \times 2 \times 1}{2 \times 1 \times 3 \times 2 \times 1} = 10.$$

More generally with n things, a of one kind and $n - a$ of another, the number of arrangements is

$$\binom{n}{a} = \frac{n!}{a!(n-a)!}.$$

In our problem the number of ways to choose 13 cards is

$$\binom{52}{13} = \frac{52!}{13!39!}.$$

The number of ways to get 13 spades is $\binom{13}{13} = \frac{13!}{13!0!} = 1$, because $0! = 1$. We multiply by 4 because of the 4 suits, and the final probability is $4 \times 13!39!/52!$, as we already found.

Binomial coefficients are discussed in PWSA, pp. 33–39.

9. Craps

The game of craps, played with two dice, is one of America's fastest and most popular gambling games. Calculating the odds associated with it is an instructive exercise.

The rules are these. Only totals for the two dice count. The player throws the dice and wins at once if the total for the first throw is 7 or 11, loses at once if it is 2, 3, or 12. Any other throw is called his "point." If the first throw is a point, the player throws the dice repeatedly until he either wins by throwing his point again or loses by throwing 7. What is the player's chance to win?

Solution for Craps

The game is surprisingly close to even, as we shall see, but slightly to the player's disadvantage.

Let us first get the probabilities for the totals on the two dice. Regard the dice as distinguishable, say red and green. Then there are $6 \times 6 = 36$ possible equally likely throws whose totals are shown in the table (next page).

By counting the cells in the table we get the probability distribution of the totals:

Total	2	3	4	5	6	7	8	9	10	11	12
P(total)	$\frac{1}{36}$	$\frac{2}{36}$	$\frac{3}{36}$	$\frac{4}{36}$	$\frac{5}{36}$	$\frac{6}{36}$	$\frac{5}{36}$	$\frac{4}{36}$	$\frac{3}{36}$	$\frac{2}{36}$	$\frac{1}{36}$

Here P means "probability of."

Cell entries give totals for game of craps

Throw of green die

Throw of red die		1	2	3	4	5	6
	1	2	3	4	5	6	7
	2	3	4	5	6	7	8
	3	4	5	6	7	8	9
	4	5	6	7	8	9	10
	5	6	7	8	9	10	11
	6	7	8	9	10	11	12

Thus the probability of a win on the first throw is

$$P(7) + P(11) = \tfrac{6}{36} + \tfrac{2}{36} = \tfrac{8}{36}.$$

The probability of a loss on the first throw is

$$P(2) + P(3) + P(12) = \tfrac{1}{36} + \tfrac{2}{36} + \tfrac{1}{36} = \tfrac{4}{36}.$$

For later throws we need the probability of making the point. Since no throws except either the point or 7 matter, we can compute for each of these the conditional probability of making the point given that it has been thrown initially. Sometimes such an approach is called the method of reduced sample spaces because, although the actual tosses produce the totals 2 through 12, we ignore all but the point and 7.

For example, for four as the point, there are 3 ways to make the point and 6 ways to make a seven, and so the probability of making the point is $3/(3 + 6) = 3/9$.

Similarly, we get the conditional probabilities for the other points and summarize:

$$4: \quad \frac{3}{3+6} = \frac{3}{9} \qquad 8: \quad \frac{5}{5+6} = \frac{5}{11}$$

$$5: \quad \frac{4}{4+6} = \frac{4}{10} \qquad 9: \quad \frac{4}{4+6} = \frac{4}{10}$$

$$6: \quad \frac{5}{5+6} = \frac{5}{11} \qquad 10: \quad \frac{3}{3+6} = \frac{3}{9}$$

Each probability of winning must be weighted by the probability of throwing the point on the initial throw to give the unconditional probability of winning for that point. Then we sum to get for the probability of winning by throwing a point

$$\tfrac{3}{36}(\tfrac{3}{9}) + \tfrac{4}{36}(\tfrac{4}{10}) + \tfrac{5}{36}(\tfrac{5}{11}) + \tfrac{5}{36}(\tfrac{5}{11}) + \tfrac{4}{36}(\tfrac{4}{10}) + \tfrac{3}{36}(\tfrac{3}{9}) \approx 0.27071.$$

To this we add the probability of winning on the first throw, $\frac{8}{36} \approx 0.22222$, to get 0.49293 as the player's probability of winning. His expected loss per unit stake is $0.50707 - 0.49293 = 0.01414$, or 1.41%. I believe that this is the most nearly even of house gambling games that have no strategy. And 1.4% doesn't sound like much, but as I write, the stock of General Motors is selling at 71, and their dividend for the year (before extras) is quoted as $2, or about 2.8%. So per two plays at craps your loss is at a rate equal to the yearly dividend payout by America's largest corporation.

Some readers may not be satisfied with the conditional probability approach used for points and may wish to see the series summed.

Let the probability of throwing the point be P and let the probability of a toss that does not count be $R(=1 - P - \frac{1}{6})$. The $\frac{1}{6}$ is the probability of throwing 7. The player can win by throwing a number of tosses that do not count and then throwing his point. The probability that he makes his point in the $(r + 1)$st throw (after the initial throw) is R^rP, $r = 0, 1, 2, \ldots$. To get the total probability, we sum over the values of r:

$$P + RP + R^2P + \cdots = P(1 + R + R^2 + \cdots).$$

Summing this infinite geometric series gives

$$\text{Probability of making point} = P/(1 - R).$$

For example, if the point is 4, $P = \frac{3}{36}$, $R = 1 - \frac{3}{36} - \frac{6}{36} = \frac{27}{36}$, $1 - R = \frac{9}{36}$, $P(\text{making the point 4}) = (3/36)/(9/36) = 3/9$, as we got by the simpler approach of reduced sample spaces.

The first time I met this problem, I summed the series and was quite pleased with myself until a few days later the reduced sample space approach occurred to me and left me deflated.

10. An Experiment in Personal Taste for Money

(a) An urn contains 10 black balls and 10 white balls, identical except for color. You choose "black" or "white." One ball is drawn at random, and if its color matches your choice, you get $10, otherwise nothing. Write down the maximum amount you are willing to pay to play the game. The game will be played just once.

(b) A friend of yours has available many black and many white balls, and he puts black and white balls into the urn to suit himself. You choose "black" or "white." A ball is drawn randomly from this urn. Write down the maximum amount you are willing to pay to play this game. The game will be played just once.

Discussion for An Experiment in Personal Taste for Money

No one can say what amount is appropriate for you to pay for either game. Even though your expected value in the first game is $5, you may not be willing to pay anything near $5 to play it. The loss of $3 or $4 may mean too much to you. Let us suppose you decided to offer 75¢.

What we *can* say is that you should be willing to pay at least as much to play the second game as the first. You can always choose your own color at random by the toss of a coin and thus assure that you have a fifty-fifty chance of being right and therefore an expectation of $5. Furthermore, if you have any information about your friend's preferences, you can take advantage of that to improve your chances.

Most people feel that they would rather play the first game because the conditions of the second seem more vague. I am indebted to Howard Raiffa for this problem, and he tells me that the idea was suggested to him by Daniel Ellsberg.

11. Silent Cooperation

Two strangers are separately asked to choose one of the positive whole numbers and advised that if they both choose the same number, they both get a prize. If you were one of these people, what number would you choose?

Discussion for Silent Cooperation

I have not met anyone yet who would choose more than a one-digit number; and of these only 1, 3, and 7 have been chosen. Most of my informants choose 1, which seems on the face of it to be the natural choice. But 3 and 7 are popular choices.

12. Quo Vadis?

Two strangers who have a private recognition signal agree to meet on a certain Thursday at 12 noon in New York City, a town familiar to neither, to discuss an important business deal, but later they discover that they have not chosen a meeting place, and neither can reach the other because both have embarked on trips. If they try nevertheless to meet, where should they go?

Discussion for Quo Vadis?

My daughter when asked this question said enthusiastically "Why, they should meet in the most famous place in New York!" "Fine," I said, "where?" "How would I know that?" she said, "I'm only 9 years old."

Places that come to mind in 1964 are top of the Empire State Building, airports, information desks at railroad stations, Statue of Liberty, Times Square. The Statue of Liberty will be eliminated the moment the strangers find out how hard it is to get there. Airports suffer from distance from town and numerosity. That there are two important railroad stations seems to me to remove them from the competition. That leaves the Empire State Building or Times Square. I would opt for the Empire State Building, because Times Square is getting vaguely large these days. I think their problem would have been easier if they had been meeting in San Francisco or Paris, don't you?

13. The Prisoner's Dilemma

Three prisoners, *A*, *B*, and *C*, with apparently equally good records have applied for parole. The parole board has decided to release two of the three, and the prisoners know this but not which two. A warder friend of prisoner *A* knows who are to be released. Prisoner *A* realizes that it would be unethical to ask the warder if he, *A*, is to be released, but thinks of asking for the name of *one* prisoner *other than himself* who is to be released. He thinks that before he asks, his chances of release are $\frac{2}{3}$. He thinks that if the warder says "*B* will be released," his own chances have now gone down to $\frac{1}{2}$, because either *A* and *B* or *B* and *C* are to be released. And so *A* decides not to reduce his chances by asking. However, *A* is mistaken in his calculations. Explain.

Solution for The Prisoner's Dilemma

Of all the problems people write me about, this one brings in the most letters.

The trouble with *A*'s argument is that he has not listed the possible events properly. In technical jargon he does not have the correct sample space. He thinks his experiment has three possible outcomes: the released pairs *AB*, *AC*, *BC* with equal probabilities of $\frac{1}{3}$. From his point of view, that is the correct sample space for the experiment conducted by the parole board given that they are to release two of the three. But *A*'s own experiment adds an event—the response of the warder. The outcomes of his proposed experiment and reasonable probabilities for them are:

1. *A* and *B* released and warder says *B*, probability $\frac{1}{3}$.
2. *A* and *C* released and warder says *C*, probability $\frac{1}{3}$.
3. *B* and *C* released and warder says *B*, probability $\frac{1}{6}$.
4. *B* and *C* released and warder says *C*, probability $\frac{1}{6}$.

If, in response to *A*'s question, the warder says "*B* will be released," then the probability for *A*'s release is the probability from outcome 1 divided by the sum of the probabilities from outcomes 1 and 3. Thus the final probability of A's release is $\frac{1}{3}/(\frac{1}{3}+\frac{1}{6})$, or $\frac{2}{3}$, and mathematics comes round to common sense after all.

14. Collecting Coupons

Coupons in cereal boxes are numbered 1 to 5, and a set of one of each is required for a prize. With one coupon per box, how many boxes on the average are required to make a complete set?

Solution for Collecting Coupons

We get one of the numbers in the first box. Now the chance of getting a new number from the next box is $\frac{4}{5}$. Using the result of Problem 4, the second new number requires $1/(4/5) = \frac{5}{4}$ boxes. The third new number requires an additional $1/(3/5) = \frac{5}{3}$; the fourth $\frac{5}{2}$, the fifth $\frac{5}{1}$.

Thus the average number of boxes required is

$$5(\tfrac{1}{5}+\tfrac{1}{4}+\tfrac{1}{3}+\tfrac{1}{2}+1) \approx 11.42.$$

Euler's Approximation for Harmonic Sums

Though it is easy to add up the reciprocals here, had there been a large number of coupons in a set, it might be convenient to know Euler's approximation for the partial sum of the harmonic series:

$$1+\frac{1}{2}+\frac{1}{3}+\cdots+\frac{1}{n} \approx \log_e n + \frac{1}{2n} + 0.57721\ldots.$$

(The 0.57721 . . . is known as Euler's constant.) For n coupons in a set, the average number of boxes is approximately

$$n \log_e n + 0.577\, n + \tfrac{1}{2}.$$

Since $\log_e 5 \approx 1.6094$, Euler's approximation for $n = 5$ yields 11.43, very close to 11.42. Often we omit the term $1/2n$ in Euler's approximation.

15. The Theater Row

Eight eligible bachelors and seven beautiful models happen randomly to have purchased single seats in the same 15-seat row of a theater. On the average, how many pairs of adjacent seats are ticketed for marriageable couples?

Solution to The Theater Row

The sequence might be (B for bachelor, M for model)

$$B\,B\,M\,M\,B\,B\,M\,B\,M\,B\,M\,B\,B\,M\,M,$$

and then 9 BM or MB pairs occur. We want the average number of unlike adjacent pairs. To be unlike, we must have BM or MB. Look at the first two positions. If they are unlike, we score one marriageable couple, if alike, we score zero. The chance of a marriageable couple in the first two seats is

$$\tfrac{8}{15} \cdot \tfrac{7}{14} + \tfrac{7}{15} \cdot \tfrac{8}{14} = \tfrac{8}{15}.$$

Furthermore $\frac{8}{15}$ is also the expected number of marriageable couples in the first two seats because $\frac{8}{15}(1) + \frac{7}{15}(0) = \frac{8}{15}$. This same calculation applies to any adjacent pair. To get the average number of marriageable adjacent pairs, we multiply by the number of adjacent pairs, 14, and get $7\frac{7}{15}$ as the expected number.

More generally, with b elements of one kind and m of another, randomly arranged in a line, the expected number of unlike adjacent elements is

$$(m + b - 1)\left[\frac{bm}{(m + b)(m + b - 1)} + \frac{mb}{(m + b)(m + b - 1)}\right] = \frac{2mb}{m + b}.$$

In our example $b = 8$, $m = 7$, giving $7\frac{7}{15}$.

The key theorem used here is that the average of a sum is the sum of the averages. We found the average number of marriageable pairs in each position, $\frac{8}{15}$ in the example, and added them up for every adjacent pair. A derivation of this theorem is given in PWSA pp. 214–216.

16. Will Second-Best Be Runner-Up?

A tennis tournament has 8 players. The number a player draws from a hat decides his first-round rung in the tournament ladder. See diagram.

Suppose that the best player always defeats the next best and that the latter always defeats all the rest. The loser of the finals gets the runner-up cup. What is the chance that the second-best player wins the runner-up cup?

Solution for Will Second-Best Be Runner-Up?

$\frac{4}{7}$. The second-best player can only get the runner-up cup if he is in the half of the ladder not occupied by the best player.

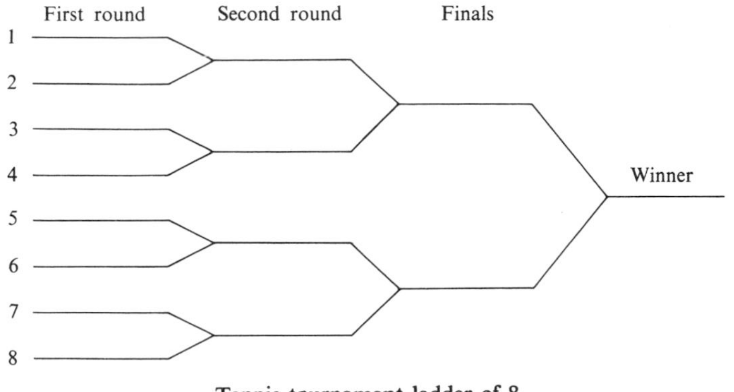

Tennis tournament ladder of 8.

In a tournament of 2^n players, there are 2^{n-1} rungs in the half (top or bottom) of the ladder not occupied by the best player, and $2^n - 1$ rungs in the whole ladder not occupied by the best player. Therefore in a tournament of 2^n players, the second-best man has probability $2^{n-1}/(2^n - 1)$ of winning the runner-up cup.

17. Twin Knights

(a) Suppose King Arthur holds a jousting tournament where the jousts are in pairs as in a tennis tournament. See Problem 16 for tournament ladder. The 8 knights in the tournament are evenly matched, and they include the twin knights Balin and Balan. What is the chance that the twins meet in a match during the tournament?

(b) Replace 8 by 2^n in the above problem. Now what is the chance that they meet?

Solution for Twin Knights

(a) Designate the twins as A and B. Put A in the top bracket (first line of the ladder). Then B is in the same bracket (pair of lines), or in the next bracket, or in the bottom half. The chance that B is adjacent to A is $\frac{1}{7}$, and then the chance they meet is 1. The chance that B is in the next pair from A is $\frac{2}{7}$, and then the chance they meet is $\frac{1}{4}$, because, to meet, each must win his first match. Finally, the chance that B is in the bottom half is $\frac{4}{7}$, and then their chance to meet is $1/2^4 = \frac{1}{16}$ because both must win 2 matches. Thus the total probability of their meeting is

$$\tfrac{1}{7} \cdot 1 + \tfrac{2}{7} \cdot \tfrac{1}{4} + \tfrac{4}{7} \cdot \tfrac{1}{16} = \tfrac{1}{4}.$$

(b) Note that for a tournament of size 2 they are sure to meet. For $2^2 = 4$ entries, their chance of meeting is $1/2$; for $2^3 = 8$ entries, we have computed their chance to be $1/4 = 1/2^2$. Thus a reasonable conjecture is that for a tournament of size 2^n, their chance of meeting is $1/2^{n-1}$.

Let us prove this conjecture by induction. We consider first the case where the knights are in opposite halves of the ladder, then the case where they are in the same half. The chance that both A and B are in opposite halves of the ladder is $2^{n-1}/(2^n - 1)$, as we know from the tennis problem immediately above. If they are in opposite halves, A and B can meet only in the finals. A knight has chance $1/2^{n-1}$ of getting to the finals because he must win $n - 1$ jousts. The chance that both A and B make the finals is $(1/2^{n-1})^2 = 1/2^{2n-2}$. Therefore the chance of their being in opposite halves and meeting is

$$[2^{n-1}/(2^n - 1)](1/2^{2n-2}).$$

To this probability must be added the chance of their being in the same half and meeting. Their chance of being in the same half is $(2^{n-1} - 1)/(2^n - 1)$, and according to the induction hypothesis, their chance of meeting in a tournament of $n - 1$ rounds is $1/2^{n-2}$. If the induction hypothesis is true, their total probability of meeting is

$$\frac{2^{n-1}}{2^n - 1} \cdot \frac{1}{2^{2n-2}} + \frac{2^{n-1} - 1}{2^n - 1} \cdot \frac{1}{2^{n-2}}$$

$$= \frac{1}{(2^n - 1)2^{n-2}} \left(\tfrac{1}{2} + 2^{n-1} - 1\right) = 1/2^{n-1},$$

which was the induction hypothesis we hoped to verify. That completes the induction.

18. An Even Split at Coin Tossing

When 100 coins are tossed, what is the probability that exactly 50 are heads?

Solution for An Even Split at Coin Tossing

Let us order the 100 coins from left to right, and then toss each one. The probability of any particular sequence of 100 tosses, a sequence of 100 heads and tails, is $(\frac{1}{2})^{100}$ because the coins are fair and the tosses independent. For example, the probability that the first 50 are heads and the second 50 are tails is $(\frac{1}{2})^{100}$. How many ways are there to arrange 50 heads and 50 tails in a row? In the Solution to the Perfect Bridge Hand (Problem 8) we found we could use binomial coefficients to make the count. We get $\binom{100}{50} = \frac{100!}{50!50!}$.

Consequently, the probability of an even split is

$$P(\text{even split}) = \frac{100!}{50!50!}\left(\frac{1}{2}\right)^{100}.$$

Evaluating this with logarithms, I get 0.07959 or about 0.08.

Stirling's Approximation

Sometimes, to work theoretically with large factorials, we use Stirling's approximation

$$n! \approx \sqrt{2\pi}\, n^{n+\frac{1}{2}}e^{-n},$$

where e is the base of the natural logarithms. The percentage error in the approximation is about $100/12n$. Let us use Stirling's approximation on the probability of an even split

$$P(\text{even split}) \approx \frac{\sqrt{2\pi}\, 100^{100+\frac{1}{2}}e^{-100}}{(\sqrt{2\pi}\, 50^{50+\frac{1}{2}}e^{-50})^2 2^{100}} = \frac{100^{\,100+\frac{1}{2}}}{\sqrt{2\pi}\, 50^{100}50(2^{100})}$$

$$= \frac{\sqrt{100}}{\sqrt{2\pi}\, 50} = \frac{1}{\sqrt{50\pi}} = \frac{1}{5\sqrt{2\pi}}.$$

Since $1/\sqrt{2\pi}$ is about 0.4, the approximation gives about 0.08 as we got before. More precisely the approximation gives to four decimals 0.0798 instead of 0.0796.

Stirling's approximation is discussed in advanced calculus books. For one nice treatment see R. Courant, *Differential and integral calculus*, Vol. I, Translated by E. J. McShane, Interscience Publishers, Inc., New York, 1937, pp. 361–364.

19. Isaac Newton Helps Samuel Pepys

Pepys wrote Newton to ask which of three events is more likely: that a person get (a) at least 1 six when 6 dice are rolled, (b) at least 2 sixes when 12 dice are rolled, or (c) at least 3 sixes when 18 dice are rolled. What is the answer?

Solution for Isaac Newton Helps Samuel Pepys

Yes, Samuel Pepys wrote Isaac Newton a long, complicated letter about a wager he planned to make. To decide which option was the favorable one, Pepys needed the answer to the above question. You may wish to read the correspondence in *American Statistician*, Vol. 14, No. 4, Oct., 1960, pp. 27–30, "Samuel Pepys, Isaac Newton, and Probability," discussion by Emil D. Schell in "Questions and Answers," edited by Ernest Rubin; and further comment in the issue of Feb., 1961, Vol. 15, No. 1, p. 29. As far as I know this is Newton's only venture into probability.

Since 1 is the average or mean number of sixes when 6 dice are thrown, 2 the average number for 12 dice, and 3 the average number for 18, one might think that the probabilities of the three events must be equal. And many would think it equal to $\frac{1}{2}$. That thought would be another instance of confusion between averages and probabilities. When the number of dice thrown is very large, then the probability that the number of sixes equals or exceeds the expected number is slightly larger than $\frac{1}{2}$. Thus for large numbers of dice, the supposition is nearly true, but not for small numbers. For large numbers of dice, the distribution of the number of sixes is approximately symmetrical about the mean, and the term at the mean is small, but for small numbers of dice, the distribution is asymmetrical and the probability of rolling exactly the mean number is substantial.

Let us begin by computing the probability of getting exactly 1 six when 6 dice are rolled. The chance of getting 1 six and 5 other outcomes in a particular order is $(\frac{1}{6})(\frac{5}{6})^5$. We need to multiply by the number of orders for 1 six and 5 non-sixes. In An Even Split at Coin Tossing, Problem 18, we learned to count the number of orders and we get $\binom{6}{1}$. Therefore the probability of exactly 1 six is

$$\binom{6}{1}\left(\frac{1}{6}\right)\left(\frac{5}{6}\right)^5.$$

Similarly, the probability of exactly x sixes when 6 dice are thrown is

$$\binom{6}{x}\left(\frac{1}{6}\right)^x\left(\frac{5}{6}\right)^{6-x}, \qquad x = 0, 1, 2, 3, 4, 5, 6.$$

The probability of x sixes for n dice is

$$\binom{n}{x}\left(\frac{1}{6}\right)^x\left(\frac{5}{6}\right)^{n-x}, \qquad x = 0, 1, \ldots, n.$$

This formula gives the terms of what is called a binomial distribution.

The probability of 1 or more sixes with 6 dice is the complement of the probability of 0 sixes:

$$1 - \binom{6}{0}\left(\frac{1}{6}\right)^0\left(\frac{5}{6}\right)^6 \approx 0.665.$$

When $6n$ dice are rolled, the probability of n or more sixes is

$$\sum_{x=n}^{6n}\binom{6n}{x}\left(\frac{1}{6}\right)^x\left(\frac{5}{6}\right)^{6n-x} = 1 - \sum_{x=0}^{n-1}\binom{6n}{x}\left(\frac{1}{6}\right)^x\left(\frac{5}{6}\right)^{6n-x}.$$

Unfortunately, Newton had to work the probabilities out by hand, but we can use the *Tables of the cumulative binomial distribution*, Harvard University Press, 1955. Fortunately, this table gives the cumulative binomial for various values of p (the probability of success on a single trial), and one of the tabled values is $p = \frac{1}{6}$. Our short table shows the probabilities, rounded to three decimals, of obtaining the mean number or more sixes when $6n$ dice are tossed.

$6n$	n	$P(n$ or more sixes)
6	1	0.665
12	2	0.619
18	3	0.597
24	4	0.584
30	5	0.576
96	16	0.542
600	100	0.517
900	150	0.514

Clearly Pepys will do better with the 6-dice wager than with 12 or 18. When he found that out, he decided to welch on his original bet.

The binomial distribution is treated extensively in PWSA, Chapter 7, see especially pp. 241–257.

20. The Three-Cornered Duel

A, *B*, and *C* are to fight a three-cornered pistol duel. All know that *A*'s chance of hitting his target is 0.3, *C*'s is 0.5, and *B* never misses. They are to fire at their choice of target in succession in the order *A*, *B*, *C*, cyclically (but a hit man loses further turns and is no longer shot at) until only one man is left unhit. What should *A*'s strategy be?

Solution for The Three-Cornered Duel

A naturally is not feeling cheery about this enterprise. Having the first shot he sees that, if he hits *C*, *B* will then surely hit him, and so he is not going to shoot at *C*. If he shoots at *B* and misses him, then *B* clearly shoots the more dangerous *C* first, and *A* gets one shot at *B* with probability 0.3 of succeeding. If he misses this time, the less said the better. On the other hand, suppose *A* hits *B*. Then *C* and *A* shoot alternately until one hits. *A*'s chance of winning is

$$(.5)(.3) + (.5)^2(.7)(.3) + (.5)^3(.7)^2(.3) + \cdots.$$

Each term corresponds to a sequence of misses by both *C* and *A* ending with a final hit by A. Summing the geometric series, we get

$$(.5)(.3)\{1 + (.5)(.7) + [(.5)(.7)]^2 + \cdots\}$$
$$= \frac{(.5)(.3)}{1 - (.5)(.7)} = \frac{.15}{.65} = \frac{3}{13} < \frac{3}{10}.$$

Thus hitting B and finishing off with C has less probability of winning for A than just missing the first shot. So A fires his first shot into the ground and then tries to hit B with his next shot. C is out of luck.

In discussing this with Thomas Lehrer, I raised the question whether that was an honorable solution under the code duello. Lehrer replied that the honor involved in three-cornered duels has never been established, and so we are on safe ground to allow A a deliberate miss.

21. Should You Sample with or without Replacement?

Two urns contain red and black balls, all alike except for color. Urn A has 2 reds and 1 black, and Urn B has 101 reds and 100 blacks. An urn is chosen at random, and you win a prize if you correctly name the urn on the basis of the evidence of two balls drawn from it. After the first ball is drawn and its color reported, you can decide whether or not the ball shall be replaced before the second drawing. How do you order the second drawing, and how do you decide on the urn?

Solution for Should You Sample with or without Replacement?

If the first ball drawn is a red, then no matter which urn is being drawn from, it now has half red and half black balls, and the second ball provides no discrimination. Therefore if red is drawn first, replace it before drawing again. If black is drawn, do not replace it. When this strategy is followed, the probabilities associated with the outcomes are

	Urn A	Urn B	decide
2 reds	$\frac{1}{2}\cdot\frac{2}{3}\cdot\frac{2}{3}$	$\frac{1}{2}\cdot\frac{101}{201}\cdot\frac{101}{201}\approx\frac{1}{8}$	Urn A
red, then black	$\frac{1}{2}\cdot\frac{2}{3}\cdot\frac{1}{3}$	$\frac{1}{2}\cdot\frac{101}{201}\cdot\frac{100}{201}\approx\frac{1}{8}$	Urn B
black, then red	$\frac{1}{2}\cdot\frac{1}{3}\cdot 1$	$\frac{1}{2}\cdot\frac{100}{201}\cdot\frac{101}{200}\approx\frac{1}{8}$	Urn A
2 black	$\frac{1}{2}\cdot\frac{1}{3}\cdot 0$	$\frac{1}{2}\cdot\frac{100}{201}\cdot\frac{99}{200}\approx\frac{1}{8}$	Urn B

The total probability of deciding correctly is approximately (replacing $\frac{100}{201}$ by $\frac{1}{2}$, etc.)

$$\tfrac{1}{2}[\tfrac{4}{9} + \tfrac{1}{4} + \tfrac{1}{3} + \tfrac{1}{4}] = \tfrac{23}{36} \approx 0.64.$$

Drawing both balls *without* replacement gives about 5/8, drawing both *with* replacement gives about 21.5/36.

22. The Ballot Box

In an election, two candidates, Albert and Benjamin, have in a ballot box a and b votes respectively, $a > b$, for example, 3 and 2. If ballots are randomly drawn and tallied, what is the chance that at least once after the first tally the candidates have the same number of tallies?

Solution for The Ballot Box

For $a = 3$ and $b = 2$, the equally likely sequences of drawings are

$A\ A\ A\ B\ B$	$*A\ A\ B\ B\ A$	$*A\ B\ B\ A\ A$
$A\ A\ B\ A\ B$	$*A\ B\ A\ B\ A$	$*B\ A\ B\ A\ A$
$*A\ B\ A\ A\ B$	$*B\ A\ A\ B\ A$	$*B\ B\ A\ A\ A$
$*B\ A\ A\ A\ B$		

where the starred sequences lead to ties, and thus the probability of a tie in this example is $\frac{8}{10}$.

More generally, we want the proportion of the possible tallying sequences that produce at least one tie. Consider those sequences in which the *first* tie appears when exactly $2n$ ballots have been counted $n \leq b$. For every sequence in which A (for Albert) is always ahead until the tie, there is a corresponding sequence in which B (for Benjamin) is always ahead until the tie. For example, if $n = 4$, corresponding to the sequence

$$A\ A\ B\ A\ B\ A\ B\ B$$

in which A leads until the tie, there is the complementary sequence

$$B\ B\ A\ B\ A\ B\ A\ A$$

in which B always leads. This second sequence is obtained from the first by replacing each A by a B and each B by an A.

Given a tie sometime, there is a first one. The number of sequences with A ahead until the first tie is the same as the number with B ahead until the first tie. The trick is to compute the probability of getting a first tie with B ahead until then.

Since A has more votes than B, A must ultimately be ahead. If the first ballot is a B, then there must be a tie sooner or later; and the only way to get a first tie with B leading at first is for B to receive the first tally. The probability that the first ballot is a B is just

$$\frac{b}{a+b}.$$

But there are just as many tie sequences resulting from the first ballot's being an A. Thus the probability of a tie is exactly

$$P(\text{tie}) = \frac{2b}{a+b} = \frac{2}{r+1},$$

where $r = a/b$. We note that when a is much larger than b, that is, when r gets large, the probability of a tie tends to zero (a result that is intuitively reasonable). And the formula holds when $b = a$, because we must have a tie and the formula gives unity as the probability.

23. Ties in Matching Pennies

Players A and B match pennies N times. They keep a tally of their gains and losses. After the first toss, what is the chance that at no time during the game will they be even?

Solution for Ties in Matching Pennies

Below we extend the method described in the Solution for The Ballot Box, Problem 22, to show that the probability of not getting a tie is (for N odd and N even)

$$P(\text{no tie}) = \binom{N-1}{n} \Big/ 2^{N-1}, \qquad N = 2n+1,$$

$$P(\text{no tie}) = \binom{N}{n} \Big/ 2^{N}, \qquad N = 2n.$$

The formulas show that the probability is the same for an even N and for the following odd number $N + 1$. For example, when $N = 4$, the second formula applies. The 16 possible outcomes are

*A A A A	B A A A	A B B A	B A B B
*A A A B	A A B B	B A B A	*B B A B
*A A B A	A B A B	B B A A	*B B B A
A B A A	B A A B	A B B B	*B B B B

where the star indicates that no tie occurs. Since the number of combinations of 4 things taken 2 at a time is 6, the formula checks.

For $N = 2n$, the probability of x wins for A is $\binom{N}{x}\Big/2^N$. If $x \leq n$, the probability of a tie is $2x/N$, based on the ballot box result, and for $x \geq n$ it is $2(N - x)/N$. To get the unconditional probability of a tie, we weight the probability of the outcome x by the probability of a tie with x wins and sum to get

$$(1) \quad 2(2^{-N})\left[\frac{0}{N}\binom{N}{0} + \frac{1}{N}\binom{N}{1} + \cdots + \frac{n-1}{N}\binom{N}{n-1} + \frac{n}{N}\binom{N}{n} + \frac{n-1}{N}\binom{N}{n+1} + \cdots + \frac{1}{N}\binom{N}{N-1} + \frac{0}{N}\binom{N}{N}\right].$$

When the binomial coefficients are converted to factorials and their coefficients canceled, we find that, except for a missing term which is $(N - 1)!/n!(n - 1)! = \binom{N-1}{n}$, the sum in brackets would be $\sum\binom{N-1}{x}$ over the possible values of x. Consequently, we can rewrite expression (1) as

$$(2) \qquad 2^{-N+1}\left[2^{N-1} - \binom{N-1}{n}\right] = 1 - \binom{N-1}{n}\Big/2^{N-1}.$$

The complement of expression (2) gives at last the probability of no tie $\binom{N-1}{n}\Big/2^{N-1}$, which a little algebra shows can be written $\binom{N}{n}\Big/2^N$ as suggested earlier.

24. The Unfair Subway

Marvin gets off work at random times between 3 and 5 P.M. His mother lives uptown, his girl friend downtown. He takes the first subway that comes in either direction and eats dinner with the one he is first delivered to. His mother complains that he never comes to see her, but he says she has a 50-50 chance. He has had dinner with her twice in the last 20 working days. Explain.

Solution for The Unfair Subway

Downtown trains run past Marvin's stop at, say, 3:00, 3:10, 3:20, . . . , etc., and uptown trains at 3:01, 3:11, 3:21, ... To go uptown Marvin must arrive in the 1-minute interval between a downtown and an uptown train.

25. Lengths of Random Chords

If a chord is selected at random on a fixed circle what is the probability that its length exceeds the radius of the circle?

Some Plausible Solutions for Lengths of Random Chords

Until the expression "at random" is made more specific, the question does not have a definite answer. The three following plausible assumptions, together with their three different probabilities, illustrate the uncertainty in the notion of "at random" often encountered in geometrical probability problems.

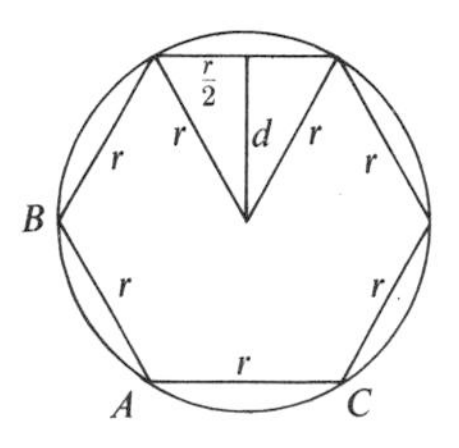

We cannot guarantee that any of these results would agree with those obtained from some physical process which the reader might use to pick random chords, indeed, the reader may enjoy studying empirically whether any do agree.

Let the radius of the circle be r.

(a) Assume that the distance of the chord from the center of the circle is evenly (uniformly) distributed from 0 to r. Since a regular hexagon of side r can be inscribed in a circle, to get the probability, merely find the distance d from the center and divide by the radius. Note that d is the altitude of an equilateral triangle of side r. Therefore from plane geometry we get $d = \sqrt{r^2 - r^2/4} = r\sqrt{3}/2$. Consequently, the desired probability is

$$r\sqrt{3}/2r = \sqrt{3}/2 \approx 0.866.$$

(b) Assume that the midpoint of the chord is evenly distributed over the interior of the circle. Consulting the figure again, we see that the chord is longer than the radius when the midpoint of the chord is within d of the center. Thus all points in the circle of radius d, concentric with the original circle, can serve as midpoints of the chord. Their fraction, relative to the area of the original circle, is $\pi d^2/\pi r^2 = d^2/r^2 = \frac{3}{4} = 0.75$. This probability is the square of the result we got from assumption (a) above.

(c) Assume that the chord is determined by two points chosen so that their positions are independently evenly distributed over the circumference of the original circle. Suppose the first point falls at A in the figure. Then for the chord to be shorter than the radius, the second point must fall on the arc BAC, whose length is $\frac{1}{3}$ the circumference. Consequently, the probability that the chord is longer than the radius is $1 - \frac{1}{3} = \frac{2}{3}$.

26. The Hurried Duelers

Duels in the town of Discretion are rarely fatal. There, each contestant comes at a random moment between 5 A.M. and 6 A.M. on the appointed day and leaves exactly 5 minutes later, honor served, unless his opponent arrives within the time interval and then they fight. What fraction of duels lead to violence?

Solution for The Hurried Duelers

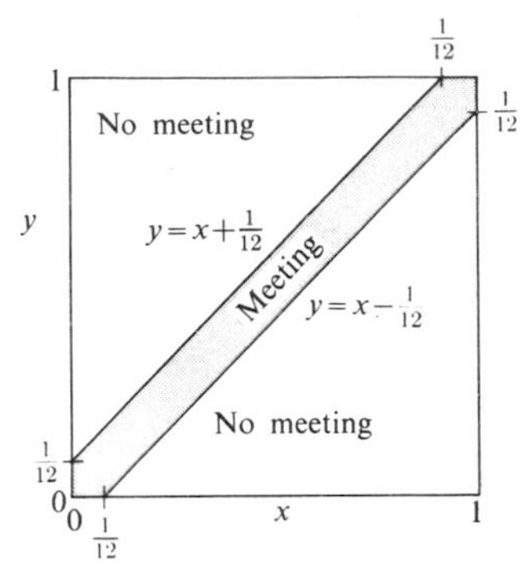

Let x and y be the times of arrivals measured in parts of an hour from 5 A.M. The shaded region of the figure shows the arrival times for which the duelists meet.

The probability that they do not meet is $(\frac{11}{12})^2$, and so the fraction of duels in which they meet is $\frac{23}{144} \approx \frac{1}{6}$.

27. Catching the Cautious Counterfeiter

(a) The king's minter boxes his coins 100 to a box. In each box he puts 1 false coin. The king suspects the minter and from each of 100 boxes draws a random coin and has it tested. What is the chance the minter's peculations go undetected?

(b) What if both 100's are replaced by n?

Solution for Catching the Cautious Counterfeiter

(a) $P(\text{0 false coins}) = (1 - \frac{1}{100})^{100} \approx 0.366$.

(b) Let there be n boxes and n coins per box.

For any box the chance that the coin drawn is good is $1 - 1/n$, and since there are n boxes,

$$P\text{ (0 false coins)} = \left(1 - \frac{1}{n}\right)^n.$$

Let us look at this probability for a few values of n.

n	P(0 false coins)
1	0
2	0.250
3	0.296
4	0.316
5	0.328
10	0.349
20	0.358
100	0.366
1000	0.3677
∞	$0.367879\ldots = 1/e$

Two things stand out. First, the tabled numbers increase; and second, they may be approaching some number. The number they are approaching is well known, and it is e^{-1} or $1/e$, where e is the base of the natural logarithms, $2.71828\ldots$.

If we expand $\left(1 - \frac{1}{n}\right)^n$ in powers of $1/n$, we get

$$1^n - \binom{n}{1} 1^{n-1}\left(\frac{1}{n}\right) + \binom{n}{2} 1^{n-2}\left(\frac{1}{n}\right)^2 - \binom{n}{3} 1^{n-3}\left(\frac{1}{n}\right)^3 + \cdots$$

or

$$(1) \qquad 1 - \frac{n}{n} + \frac{n(n-1)}{2!n^2} - \frac{n(n-1)(n-2)}{3!n^3} + \cdots.$$

If we take one of these terms, say the fourth, and study its behavior as n becomes very large, we find that it approaches $-1/3!$ because

$$(2) \qquad \frac{n(n-1)(n-2)}{n^3} = 1\left(1 - \frac{1}{n}\right)\left(1 - \frac{2}{n}\right) = 1 - \frac{3}{n} + \frac{2}{n^2}.$$

As n grows large, all terms on the right-hand side of eq. (2) except the 1 tend to zero. Similarly, for the rth term of expansion (1) the factors depending on n tend to 1, and the term itself tends except for sign to $1/(r-1)!$. Therefore, as n grows, the series for $\left(1 - \frac{1}{n}\right)^n$ tends to

$$1 - 1 + \frac{1}{2!} - \frac{1}{3!} + \frac{1}{4!} - \frac{1}{5!} + \cdots.$$

This series is one way of writing e^{-1}.

Had we investigated the case of 2 false coins in every box, we would have found that $\left(1 - \frac{2}{n}\right)^n$ tends to e^{-2} as n grows large, and in general that $\left(1 - \frac{m}{n}\right)^n$ tends to e^{-m}. Also $\left(1 + \frac{m}{n}\right)^n$ tends to e^m whether m is an integer or not. These facts are important for us. They can be studied at more leisure and more rigorously in calculus books, for example, Thomas, G. B., Jr., *Elements of calculus and analytic geometry*, Addison-Wesley, Reading, Mass., 1959, pp. 384–399.

28. Catching the Greedy Counterfeiter

The king's minter boxes his coins n to a box. Each box contains m false coins. The king suspects the minter and randomly draws 1 coin from each of n boxes and has these tested. What is the chance that the sample of n coins contains exactly r false ones?

Solution for Catching the Greedy Counterfeiter 28

Each of the coins in the king's sample is drawn from a new box and has probability m/n of being counterfeit. The drawings are independent, and so we get the binomial probability for r false (and $n - r$ true) to be

$$P(r \text{ false coins}) = \binom{n}{r}\left(\frac{m}{n}\right)^r\left(1 - \frac{m}{n}\right)^{n-r}.$$

Let us see what happens when n grows large while r and m are fixed. We write $P(r$ false coins) as

$$\frac{1}{r!}\frac{n(n-1)\cdots(n-r+1)}{n^r}\cdot m^r\left(1-\frac{m}{n}\right)^n\left(1-\frac{m}{n}\right)^{-r}.$$

As n grows, $1/r!$ is unchanged, m^r is unchanged,

$$n(n-1)\cdots(n-r+1)/n^r$$

tends to 1, $\left(1 - \frac{m}{n}\right)^n$ tends to e^{-m}, as explained in Problem 27, and $\left(1 - \frac{m}{n}\right)^{-r}$ tends to 1 (again because m and r are fixed). Therefore for large n

$$P(r \text{ false coins}) \approx \frac{e^{-m}m^r}{r!}.$$

These terms add up to 1, that is,

$$e^{-m}\left(\frac{1}{0!} + \frac{m}{1!} + \frac{m^2}{2!} + \frac{m^3}{3!} + \cdots\right) = e^{-m}e^m = 1.$$

The series in parentheses is an expansion of e^m.

Poisson Distribution

The distribution whose probabilities are

$$P(r) = \frac{e^{-m}m^r}{r!}, \qquad r = 0, 1, 2, \ldots,$$

is called the Poisson distribution, and it approximately represents the probabilistic behavior of many physical processes.

You might read about the Poisson distribution in M. J. Moroney, *Facts from figures*, 3rd ed., Penguin Books, Ltd., Baltimore, Maryland, 1956, pp. 96–107.

29. Moldy Gelatin

Airborne spores produce tiny mold colonies on gelatin plates in a laboratory. The many plates average 3 colonies per plate. What fraction of plates has exactly 3 colonies? If the average is a large integer m, what fraction of plates has exactly m colonies?

Solution for Moldy Gelatin

Regard the surface of a plate as broken into n small equal areas. For each area the probability of a colony is p, but the mean number is $np = 3$. We want tiny areas. As n grows, p becomes small, because the area of a subregion tends to zero. Instead of fixing on a mean number of 3, let us keep a general mean, $m = np$. You may be worrying that in some areas 2 or more colonies can occur, but relax, because the little regions will be so small they can barely hold one colony. Then the probability of exactly r colonies in n small areas is the binomial

$$\binom{n}{r} p^r(1 - p)^{n-r},$$

where $p = m/n$. Replace p by m/n. Then, the formula is our old friend from the Greedy Counterfeiter, Problem 28. Let n tend to infinity, and we again get the Poisson distribution

$$P(r) = \frac{e^{-m}m^r}{r!}, \qquad r = 0, 1, 2, \ldots.$$

For $m = 3$, and $r = 3$, this yields 0.224.

You might verify from the definition that m is the mean of the distribution as follows:

$$\text{mean} = \sum_{x=0}^{\infty} xP(x) = m \sum_{x=1}^{\infty} e^{-m}m^{x-1}/(x - 1)! = m.$$

Several good tables of the Poisson are now available:

T. C. Fry, *Probability and its engineering uses*, D. Van Nostrand Company, Inc., Princeton, New Jersey, 1928, pp. 458–467.

T. Kitagawa, *Tables of Poisson distribution*, Baifukan, Tokyo, Japan, 1952.

E. C. Molina, *Poisson's exponential binomial limit*, D. Van Nostrand Company, Inc., Princeton, New Jersey, 1942.

To get the results for a large value of m, where $r = m$, we could use the tables or apply Stirling's approximation. Stirling's approximation gives

$$P(m) = \frac{e^{-m}m^m}{m!} \approx \frac{e^{-m}m^m}{\sqrt{2\pi}\, m^{m+\frac{1}{2}}e^{-m}} = \frac{1}{\sqrt{2\pi\, m}} \approx \frac{0.4}{\sqrt{m}}.$$

Numerical examples:

m	$P(m)$	$0.4/\sqrt{m}$
4	0.1954	0.200
9	0.1318	0.133
16	0.0992	0.100

30. Evening the Sales

A bread salesman sells on the average 20 cakes on a round of his route. What is the chance that he sells an even number of cakes? (We assume the sales follow the Poisson distribution.)

Solution for Evening the Sales

Why assume a Poisson? Partly because the problem is nice that way. Partly because the distribution may be close to Poisson because the salesman has many customers, each with a small chance of buying a cake. You may be worried about variation from day to day during the week—good for you—I'm thinking only of summer Tuesdays.

Most of us would guess about $\frac{1}{2}$.

The probability of his selling exactly r cakes is $e^{-20}20^r/r!$, as we know from Problem 28. Working with the general mean m instead of 20 will clarify the structure of the problem. Then, the sum of the Poisson probabilities is $\sum e^{-m}m^r/r!$, or

$$\text{(A)} \qquad 1 = e^{-m}e^{m} = e^{-m}\left(1 + \frac{m}{1!} + \frac{m^2}{2!} + \frac{m^3}{3!} + \frac{m^4}{4} + \cdots\right).$$

We want to eliminate the terms corresponding to odd numbers of cakes. Recall that

$$\text{(B)} \quad e^{-2m} = e^{-m}e^{-m} = e^{-m}\left(1 - \frac{m}{1!} + \frac{m^2}{2!} - \frac{m^3}{3!} + \frac{m^4}{4!} - \cdots\right).$$

The sum of expressions (A) and (B) would give us twice the probability of an even number of loaves because the terms with odd powers of m would add to zero and the terms with even powers would have a coefficient of 2. Consequently, after dividing by 2, we get for the probability of an even number $(1 + e^{-2m})/2$. For $m = 20$ the result is extremely close to 0.5 because e^{-40} is negligible.

On the other hand, if he sold on the average one special birthday cake per trip over the route, the probability that he sells an even number of special birthday cakes is about 0.568.

31. Birthday Pairings

What is the least number of persons required if the probability exceeds $\frac{1}{2}$ that two or more of them have the same birthday? (Year of birth need not match.)

Solution for Birthday Pairings

The usual simplifications are that February 29 is ignored as a possible birthday and that the other 365 days are regarded as equally likely birth dates.

Let us solve a somewhat more general problem. Let N be the number of equally likely days, r the number of individuals, and let us compute the probability of no like birthdays. Then we can get the probability of at least one pair of like birthdays by taking the complement.

There are N days for the first person to have a birthday, $N - 1$ for the second so that he does not match the first, $N - 2$ for the third so that he matches neither of the first two, and so on down to $N - r + 1$ for the rth person. Then apply the multiplication principle and find the number of ways for no matching birthdays to be

$$N(N-1)\cdots(N-r+1), \qquad r \text{ factors.} \tag{1}$$

To get the probability of no matching birthdays we also need the number of ways r people can have birthdays without restriction. There are N ways for each person. Then the multiplication principle says that the total number of different ways the birthdays can be assigned to r people is

$$N^r. \tag{2}$$

The number in expression (1) divided by that in expression (2) is the probability of no like birthdays, because we assume that all birthdays and therefore all ways of assigning birthdays are equally likely. The complement of this ratio is the probability of at least one pair of like birthdays.

Thus

$$\begin{aligned} P_R &= P \text{ (at least 1 matching pair)} \\ &= 1 - N(N-1)\cdots(N-r+1)/N^r. \end{aligned} \tag{3}$$

To evaluate expression (3) for large values of N such as 365 requires some courage or, better, some good tables of logarithms. T. C. Fry in *Probability and its engineering uses*, D. Van Nostrand Company, Inc., Princeton, New Jersey, 1928, gives tables of logarithms of factorials, and so it is convenient to evaluate the probability of no like birthdays in the form

$$\frac{N!}{(N-r)!\,N^r}.$$

The following data help:

$$\log 365! = 778.39975 \qquad \log 365 = 2.56229286$$

$$\begin{aligned} r &= 20, & \log 345! &= 727.38410 \\ r &= 21, & \log 344! &= 724.84628 \\ r &= 22, & \log 343! &= 722.30972 \\ r &= 23, & \log 342! &= 719.77442 \\ r &= 24, & \log 341! &= 717.24040 \\ r &= 25, & \log 340! &= 714.70764 \end{aligned}$$

A short bout with tables of logarithms shows that for $r = 23$ the probability of at least one success is 0.5073, but for $r = 22$ the probability is 0.4757. Thus $r = 23$ is the least number that gives a 50-50 chance of getting some like birthdays. Most persons are surprised that the number required is so small for they expected about 365/2. We discuss that notion in our next problem, but let us do a bit more with the current one.

First, the table gives probabilities of at least one pair of like birthdays for various values of r:

r	5	10	20	23	30	40	60
P_R	0.027	0.117	0.411	0.507	0.706	0.891	0.994

Second, let us learn a tricky way to approximate the probability of failure. Recall that

$$e^{-x} = 1 - x + \frac{x^2}{2!} - \frac{x^3}{3!} + \cdots.$$

If x were very small, then the terms beyond $1 - x$ would not amount to much. Consequently, for small values of x we might approximate e^{-x} by $1 - x$ or, as in what follows, $1 - x$ by e^{-x}. Note that $N(N-1)\cdots(N-r+1)/N^r$ is a product of factors $(N-k)/N$, where k is much smaller than N. These factors can be written as $1 - k/N$, where $0 \le k \le r$. Therefore

$$\begin{aligned} N(N-1)\cdots(N-r+1)/N^r &\approx e^{-[0+1+\cdots+(r-1)]/N} \\ &= e^{-r(r-1)/2N}. \end{aligned}$$

To see the approximation in action, try it on $r = 23$ and get about 0.500 instead of 0.507. Or set $r(r-1)/2(365)$ equal to $-\log_e 0.5 \approx 0.693$ and solve for r.

Third, suppose the original problem were extended so that you wanted the least number to achieve at least one pair of either identical birthdays or adjacent birthdays (December 31 is adjacent to January 1). Try this problem on your own.

32. Finding Your Birthmate

You want to find someone whose birthday matches yours. What is the least number of strangers whose birthdays you need to ask about to have a 50-50 chance?

Solution for Finding Your Birthmate

I think this personal birthmate problem is what most persons think of when they are asked about Birthday Pairings, Problem 31. From their notions about the personal birthmate problem stems their surprise at $r = 23$ for the previous problem. In the current birthmate problem it is of no use to you if two *other* persons have the same birthday unless it matches yours. For this problem most people reason that the number should be about half of 365 or, say, 183. Since they have confused the pairings problem with this one, they regard 23 as very small.

While good marks should be given for 183 for the birthmate problem to persons working it in their heads, even here that number is not close to the correct value because the sampling of births is done with replacement. If your first candidate is born on the Fourth of July, that does not use up the date, and later candidates may also be born on that date. Indeed, each candidate's chance to miss matching your birthday is $(N - 1)/N$, where $N = 365$, the number of days in a year. When you examine n people, the probability that none of them have your birthday is $[(N - 1)/N]^n$, and so the probability that at least one matches is

$$P_S = 1 - \left(\frac{N-1}{N}\right)^n. \tag{4}$$

We need to find the smallest n so that P_S is at least $\frac{1}{2}$. The logarithm of 364 is 2.56110, of $\frac{1}{2}$ is -0.30103.

If we solve the resulting problem in logarithms, we find that n should be 253, quite a bit more than 183.

Alternatively, we could use again the approximation

$$\frac{N-1}{N} = 1 - \frac{1}{N} \approx e^{-1/N}.$$

Then we require approximately

$$P_S \approx 1 - e^{-n/N} = \tfrac{1}{2}.$$

Consequently,

$$e^{-n/N} = \tfrac{1}{2}.$$

Taking natural logarithms gives us

$$n/N \approx 0.693,$$
$$n \approx 0.693N.$$

And for $N = 365$, $n = 253$.

This birthmate problem is easier to solve than the pairings problem, and so it would be nice to have a relation between the two answers.

33. Relating the Birthday Pairings and Birthmate Problems

If r persons compare birthdays in the pairings problem, the probability is P_R that at least 2 have the same birthday. What should n be in the personal birthmate problem to make your probability of success approximately P_R?

Solution for Relating the Birthday Pairings and Birthmate Problems

Essentially, the issue is the number of *opportunities* for paired birthdays. In the birthmate problem, n persons offer n opportunities to find your own birthmate. In the birthday-pairings problem, each individual compares his own birthday with $r - 1$ others. The number of such pairs among r persons is $r(r - 1)/2$, and that is the number of opportunities for like birthdays. To get approximately the same probability in the two problems, we should have

(1) $$n \approx r(r - 1)/2.$$

For example, when $r = 23$, n should be about $23(22)/2=253$, which agrees exactly with our findings in the two previous problems.

Those who wish to pursue this further might see "Understanding the birthday problem," *Mathematics Teacher*, Vol. 55, 1962, pp. 322–5.

In the previous two problems we found that, for n much less than N, the probability of not finding one's own birthmate among n people is approximately $e^{-n/N}$. Similarly, we found in the birthday-pairings problem that, for r small compared with N, the probability of not finding a pair with identical birthdays is approximately $e^{-r(r-1)/2N}$. For the two probabilities to be nearly equal, expression (1) must hold. This direct attack through the approximation gives us one way to understand the relation between the problems. The earlier discussion makes clear that $r(r-1)/2$ has the physical interpretation "number of opportunities" which gave another explanation for comparing n with $r(r-1)/2$.

34. Birthday Holidays

Labor laws in Erewhon require factory owners to give every worker a holiday whenever one of them has a birthday and to hire without discrimination on grounds of birthdays. Except for these holidays they work a 365-day year. The owners want to maximize the expected total number of man-days worked per year in a factory. How many workers do factories have in Erewhon?

Solution for Birthday Holidays

With 1 worker in the factory, the owner gets 364 man-days, with 2 he usually gets 2(363) = 726, and so we anticipate more than 2 workers to maximize working days in a factory. On the other hand, if the factory population is enormous, every day of the year is practically certain to be someone's birthday, and the factory never works. Consequently, there must be a finite maximum.

If we can get the expected total number of days worked, we are a long step forward. Each day is either a working day or it isn't. Let's replace 365 by N so that we solve the problem generally, and let n be the number of workers. Then the probability that the first day is a working day is $(1 - 1/N)^n$, because then every worker has to have a birthday on one of the other $N - 1$ days. The expected number of man-days contributed by the first working day is

$$n\left(1 - \frac{1}{N}\right)^n \cdot 1 + n\left[1 - \left(1 - \frac{1}{N}\right)^n\right] \cdot 0 = n\left(1 - \frac{1}{N}\right)^n.$$

Every day contributes this same number, and so the expected number of man-days worked by n workers is $nN(1 - 1/N)^n$. To maximize this function of n, we must find n so that increasing or decreasing n reduces the total, or in symbols:

$$(n + 1)N\left(1 - \frac{1}{N}\right)^{n+1} \leq nN\left(1 - \frac{1}{N}\right)^n$$

and

$$(n - 1)N\left(1 - \frac{1}{N}\right)^{n-1} \leq nN\left(1 - \frac{1}{N}\right)^n.$$

The first inequality reduces to:

$$(n + 1)\left(1 - \frac{1}{N}\right) \leq n,$$

$$N \leq n + 1.$$

The second inequality reduces to:

$$n - 1 \leq n\left(1 - \frac{1}{N}\right),$$

$$n \leq N.$$

Combining these results gives us $n \leq N \leq n + 1$, and so either $n = N$ or $n = N - 1$. When these values are substituted for n in the formula for the expected man-days, we get $N^2(1 - 1/N)^N$ and $(N - 1)N(1 - 1/N)^{N-1}$, which are equal. Since the Nth man adds nothing, $N - 1$ must be the factory size. Since $(1 - 1/N)^N \approx e^{-1}$, we get at last N^2e^{-1} as the approximate expected number of days worked. If all N men worked every day, they would work N^2 days, and so e^{-1} is the expected fraction that the actual man-days worked is of the potential N^2 man-days. Thus the fraction is about 0.37. The factory size is 364, and the man-days worked are roughly 49,000, assuming no other absenteeism. The 364th worker adds only 0.37 days to the total expectation! Labor must be very cheap in Erewhon.

35. The Cliff-Hanger

From where he stands, one step toward the cliff would send the drunken man over the edge. He takes random steps, either toward or away from the cliff. At any step his probability of taking a step away is $\frac{2}{3}$, of a step toward the cliff $\frac{1}{3}$. What is his chance of escaping the cliff?

Solution for The Cliff-Hanger

Before trying to solve a problem, I find it a help to see what is happening. Let us see what could happen in the first few steps. The diagram illustrates that the man can go over the cliff only on an odd-numbered step. After one step, he had a $\frac{1}{3}$ chance of being over the cliff. The path through the positions $1 \to 2 \to 1 \to 0$ adds another $\frac{2}{27}$ to the probability of disaster for a total of $\frac{11}{27}$. At the end of 5 steps, the paths $1 \to 2 \to 1 \to 2 \to 1 \to 0$ and $1 \to 2 \to 3 \to 2 \to 1 \to 0$ have together added $\frac{8}{243}$ to the probability of disaster for a total of $\frac{107}{243}$. One could extend the table, and one might learn something from a further analysis of the probabilities. I turn now to a different attack.

This famous random walk problem has many forms. Next, we shall treat it as a particle moving along an axis.

Consider a particle initially at position $x = 1$ on the real line. The structure of the problem will be clearer if we let p, rather than $\frac{2}{3}$, be the probability of a step to the right. The particle moves from position 1 either to position $x = 2$ with probability p or to position $x = 0$ with probability

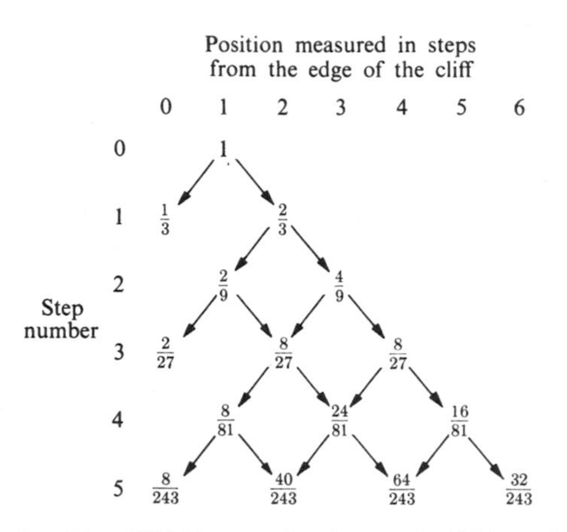

Diagram for The Cliff-Hanger showing probabilities of the man's being at various distances from the edge.

$1 - p$. More generally, if the particle is at position $x = n$, $n > 0$, n an integer, its next move is either to $x = n + 1$ with probability p or to $x = n - 1$ with probability $1 - p$. If the particle ever arrives at $x = 0$, it is absorbed there (takes no further steps). We wish to know the probability, P_1, that the particle is absorbed at $x = 0$, given that it starts at $x = 1$. Naturally, the value of P_1 depends upon p. It seems reasonable that if p is near 1, P_1 is small, but if p is near 0, P_1 is close to 1.

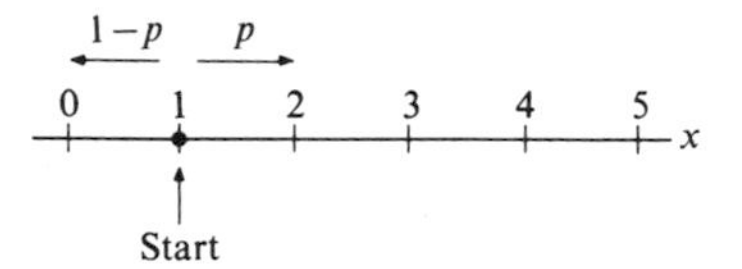

Consider the situation after the first step: either the particle moved left to $x = 0$ and was absorbed (this event has probability $1 - p$) or it moved right to $x = 2$ (this event has probability p). Let P_2 be the probability of the particle's being absorbed at $x = 0$ when the particle starts from position $x = 2$. Then we can write

$$P_1 = 1 - p + pP_2, \tag{1}$$

because $1 - p$ is the probability of absorption at the first step and pP_2 is the probability of being absorbed later.

Paths leading to absorption from $x = 2$ can be broken into two parts: (1) a path that goes from $x = 2$ to $x = 1$ for the first time (not necessarily in one step), and (2) a path from $x = 1$ to $x = 0$ (again, not necessarily in one step). The probability of a path from $x = 2$ to $x = 1$ is just P_1, *because the structure here is identical with that of the original set-up for the particle except that the origin has been translated one step to the right.* The probability of a path from $x = 1$ to $x = 0$ is also P_1, because this is exactly the original problem. The probability P_2 therefore is P_1^2, because the events A = (particle takes path from $x = 2$ to $x = 1$) and B = (particle takes path from $x = 1$ to $x = 0$) are independent, and $P(A) = P(B) = P_1$.

We can rewrite eq. (1) as

$$P_1 = 1 - p + pP_1^2. \tag{2}$$

Equation (2) is quadratic in P_1 with solutions

$$P_1 = 1, \quad P_1 = \frac{1-p}{p}. \tag{3}$$

In such problems one or both solutions may be appropriate, depending on the circumstances.

We need to choose the solution that goes with each value of p. When $p = \frac{1}{2}$, the solutions agree, and $P_1 = 1$. When $p = 0$, clearly $P_1 = 1$. And when $p = 1$, $P_1 = 0$, because the particle always moves to the right. When $p < \frac{1}{2}$, the second solution of (3) is impossible because then $(1 - p)/p > 1$, and we must have $P_1 \leq 1$. Therefore, for $0 \leq p \leq \frac{1}{2}$, we have $P_1 = 1$.

To prove that the second solution, $P_1 = (1 - p)/p$, holds for $p > \frac{1}{2}$ requires us to show that P_1 is a continuous function of p (roughly, that P_1 does not jump when p changes slightly). We assume this continuity but do not prove it.

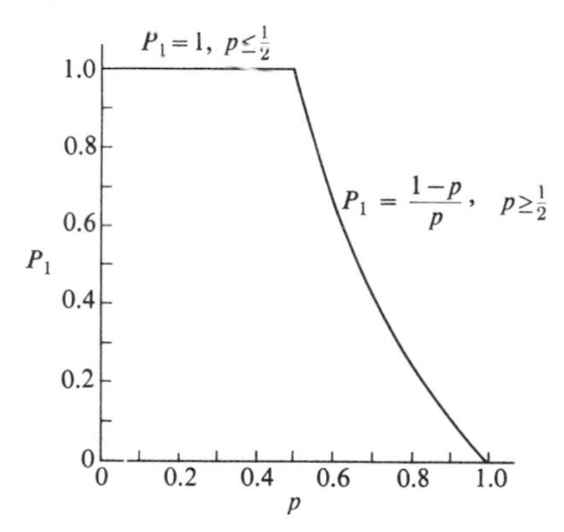

Probabilities of absorption, P_1, for The Cliff-Hanger.

The curve (see figure) starts at $P_1 = 1$ when $p = \frac{1}{2}$, must decrease to $P_1 = 0$ at $p = 1$, and must always have value either 1 or $(1 - p)/p$. For it to avoid jumps, it must adopt $(1 - p)/p$ for $p > \frac{1}{2}$. The proof of the continuity itself is beyond the scope of this book, but assuming the continuity, for $p > \frac{1}{2}$ we have $P_1 = (1 - p)/p$. Therefore our cliff-hanging man has probability $\frac{1}{2}$ of falling over the cliff.

To give another interpretation, if a gambler starting with one unit of money ($x = 1$) could and did play indefinitely against a casino with infinitely large resources a fair game ($p = \frac{1}{2}$) in which he wins or loses one unit on each play, he would be certain to lose his money ($P_1 = 1$). To have an even chance of not going bankrupt, he must have $p = \frac{2}{3}$.

That bankruptcy is certain for $p = \frac{1}{2}$ is surprising for most of us. We usually suppose that if the trials of a game are "fair" (average loss is zero), then the whole game is fair. Indeed, this supposition is ordinarily correct. If we imagine this game, with $p = \frac{1}{2}$, being played infinitely many times, then the average amount of money on hand after n plays is 1, for every finite number n. So the unfairness is one of those paradoxes of the infinite. Another surprise for $p = \frac{1}{2}$ is that the average number of trials required for absorption is not finite. The case $p = \frac{1}{2}$ is indeed strange and deep.

You may enjoy applying the technique given here to a particle that starts at $x = m$, rather than $x = 1$, and generalizing the above results to show that the probability of absorption from position m is $[(1 - p)/p]^m$ or 1 depending on whether p is greater or less than $\frac{1}{2}$. When $p > \frac{1}{2}$ and m is large, it is extremely plausible that the particle escapes and therefore we would reject 1 as the absorption probability.

Had the particle started at 0 and been allowed to take its steps in either direction with $p = \frac{1}{2}$, another classical random walk problem would ask whether the particle would ever return to the origin. We see it would because it is sure to return from $x = 1$ and from $x = -1$. More on this later.

36. Gambler's Ruin

Player M has \$1, and Player N has \$2. Each play gives one of the players \$1 from the other. Player M is enough better than Player N that he wins $\frac{2}{3}$ of the plays. They play until one is bankrupt. What is the chance that Player M wins?

Solution for Gambler's Ruin

Our problem is a special case of the general random walk problem with two absorbing barriers. Historically, the problem arose as a gambling problem, called "gambler's ruin," and many famous mathematicians have contributed to questions arising from it. Let us restate the problem generally.

Player M has m units; Player N has n units. On each play of a game one player wins and the other loses 1 unit. On each play, the probability that Player M wins is p, that N wins is $q = 1 - p$. Play continues until one player is bankrupt. The figure represents the amount of money Player M

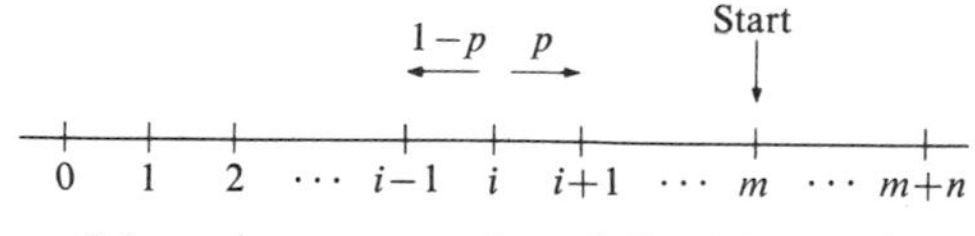

Schematic representation of Gambler's Ruin.

has at any time. He starts at $x = m$. When $x = 0$, he is bankrupt; when $x = m + n$, Player N is bankrupt.

With this representation, since $p > \frac{1}{2}$, we can appeal to a result from The Cliff-Hanger, Problem 35. We know that, had Player M played against a bank with unlimited resources, he would have become bankrupt with probability $(q/p)^m$. In the course of a trip to bankruptcy, either he attains an amount of money $m + n$ (n is now finite), or he is never that well off. Let the probability that he loses to Player N be Q (that is equivalent to the infinite bank winning without Player M ever reaching $m + n$). Then

$$(q/p)^m = Q + (1 - Q)(q/p)^{m+n}, \tag{1}$$

because Q is the fraction of the sequences that are absorbed before reaching $m + n$, and of the fraction $1 - Q$ that do reach $m + n$, the portion $(q/p)^{m+n}$ is also absorbed at 0 if the game is allowed to proceed indefinitely. Then $P = 1 - Q$ is the probability that Player M wins. Making substitutions into eq. (1) and solving for P gives

$$P = \frac{1 - (q/p)^m}{1 - (q/p)^{m+n}}. \tag{2}$$

For our players $p = \frac{2}{3}$, $q = \frac{1}{3}$, $m = 1$, $n = 2$, and $P = \frac{4}{7}$. So in this instance it is better to be twice as good a player rather than twice as wealthy.

If $q = p = \frac{1}{2}$, then P in eq. (2) takes the indeterminate form 0/0. When L'Hospital's rule is applied, we find

$$P = \frac{m}{m + n}, \quad p = q = \tfrac{1}{2}. \tag{3}$$

Thus, had the players been evenly matched, Player M's chance would be $\frac{1}{3}$ and his expectation would be $\frac{1}{3}(2) + \frac{2}{3}(-1) = 0$. Thus the game is fair, that is, has 0 expectation of gain for each player.

37. Bold Play vs. Cautious Play

At Las Vegas, a man with \$20 needs \$40, but he is too embarrassed to wire his wife for more money. He decides to invest in roulette (which he doesn't enjoy playing) and is considering two strategies: bet the \$20 on "evens" all at once and quit if he wins or loses, or bet on "evens" one dollar at a time until he has won or lost \$20. Compare the merits of the strategies.

Solution for Bold Play vs. Cautious Play

Bold play, as Lester E. Dubins and Leonard J. Savage call it in their *How to gamble if you must**, that is, betting \$20 at once, gives him a probability of $\frac{18}{38} \approx 0.474$ of achieving his goal.

Cautious play, a dollar at a time, leads us to the gambler's ruin problem with

$$m = 20, \qquad n = 20,$$

$$p = \tfrac{18}{38}, \qquad q = \tfrac{20}{38}.$$

Substituting into the formula for M's chance to win obtained in Problem 36 gives us

$$P = \frac{1 - (\frac{20}{18})^{20}}{1 - (\frac{20}{18})^{40}} = \frac{8.23 - 1}{67.7 - 1} \approx 0.11.$$

Cautious play has reduced his chances of reaching the goal to less than one-fourth of that for bold play.

The intuitive explanation is that bold play is also fast play, and fast play reduces the exposure of the money to the house's percentage. We have several times seen that intuitions based on averages do not always lead to correct probabilities. Dubins and Savage warn that no known proof of the merits of bold play, in general, is based upon this intuitive argument. However, Dubins points out that for our special case of doubling one's money at Red-and-Black, the following exposition by Savage is so based. In preparing this discussion for us, Savage has deliberately glossed over a couple of mathematical fine points dealing with the attainability of bounds.

The Golden Paradise

At the Golden Paradise they sell any fair gamble that a gambler has the funds to stake on. A gambler who enters the Golden Paradise with x dollars bent on making an income of y additional dollars, if possible, can achieve his goal with probability $x/(x + y)$ by staking his entire fortune x on a single chance of winning y with probability $x/(x + y)$, which is plainly fair. As is well known, no strategy can give him a higher probability of achieving his goal, and the probability is this high if and only if he makes sure either to lose x or win y eventually.

The Lesser Paradise

The Lesser Paradise resembles the Golden Paradise with the important difference that before leaving the hall the gambler must pay an income tax

*First published, 1965. Reprinted by Dover Publications, Inc. in 1976 under the title *Inequalities for stochastic processes.*

of t 100% ($0 < t < 1$) on any net positive income that he has won there. It is therefore no harder or easier for him to win y dollars with an initial fortune of x than it is for his brother in the Golden Paradise to win $y/(1 - t)$ dollars. The greatest probability with which he can achieve his goal is therefore

$$\frac{(1 - t)x}{(1 - t)x + y}. \tag{1}$$

The Paradise Lost

Here, the croupier collects the tax of t 100% on the positive income, if any, of each individual gamble. The gambler here is evidently in no way better off than his brother in the Lesser Paradise. In particular, (1) is an upper bound on the probability of winning y with an initial fortune of x in the Paradise Lost. This probability can be achieved by staking all on a single chance as before. However, it cannot be achieved by any strategy that has positive probability of employing some gamble that has positive probability of winning any positive amount less than y after taxes. To see this, consider that the Lesser Paradise brother can imitate any strategy of the Paradise Lost brother, setting aside for his own later use whatever the croupier takes from the Paradise Lost brother in taxes on small prizes. Thus, the former can have a higher expected income than the latter can have on any strategy in which he risks winning a small prize.

Red-and-Black

In Red-and-Black, the gambler can stake any amount in his possession against a chance of probability w ($0 < w < \frac{1}{2}$) of winning a prize equal to his stake. Put differently, he wins the fair prize of $(1 - w)/w$ times his stake subject to an immediate tax of t 100%, where

$$t = \frac{1 - 2w}{1 - w}.$$

Therefore the probability for a gambler in Red-and-Black to win y with an initial fortune of x is at most (1), as it is for his brother in Paradise Lost. In terms of w, this is

$$\frac{wx}{wx + (1 - w)y}. \tag{2}$$

Moreover, the bound (2) can be achieved only if the gambler in Red-and Black can avoid any positive probability of ever winning a positive amount less than y on his individual gambles and be sure of either losing exactly x or winning exactly y. As is not hard to see, this can occur only if $y = x$, in which case he can win y with a single bold gamble with the probability w given by (2).

The problem of an exact upper bound and optimum strategies for the gambler in Red-and-Black who wants to win an amount different from x is more difficult and will not be entered into here.

38. The Thick Coin

How thick should a coin be to have a $\frac{1}{3}$ chance of landing on edge?

Solution for The Thick Coin

On first hearing this question, the late great mathematician, John von Neumann, was unkind enough to solve it—including a 3-decimal answer—in his head in 20 seconds in the presence of some unfortunates who had labored much longer.

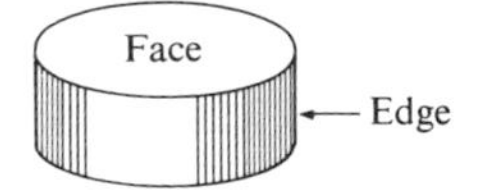

This problem has no definite answer without some simplifying conditions. The elasticity of the coin, the intensity with which it is tossed, and the properties of the surface on which it lands combine to make the real-life question an empirical one.

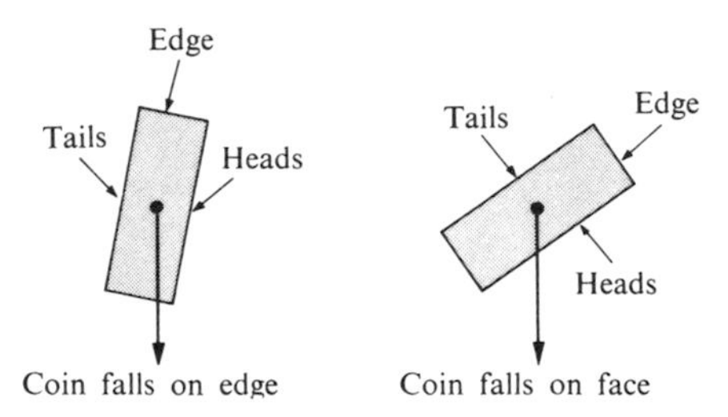

The simplifying conditions that spring to mind are those that correspond to inscribing the coin in a sphere, where the center of the coin is the center of the sphere. The coin itself is regarded as a right circular cylinder. Then a random point on the surface of the sphere is chosen. If the radius from that point to the center strikes the edge, the coin is said to have fallen on edge.

To simulate this in reality, the coin might be tossed in such a way that it fell on a thick sticky substance that would grip the coin when it touched, and then the coin would slowly settle to its edge or its face.

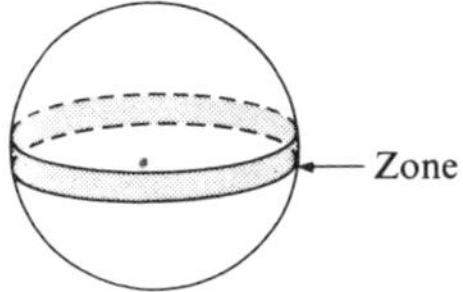

A key theorem in solid geometry simplifies this problem. When parallel planes cut a sphere, the orange-peel-like band produced between them is called a zone. The surface area of a zone is proportional to the distance between the planes, and so our coin should be $\frac{1}{3}$ as thick as the sphere. How should the thickness compare with the diameter of the coin?

Let R be the radius of the sphere and r that of the coin.

The Pythagorean theorem gives

$$R^2 = r^2 + \frac{1}{9}R^2,$$

or

$$\frac{8}{9}R^2 = r^2,$$

$$\frac{R^2}{9} = \frac{r^2}{8},$$

$$\frac{1}{3}R = \frac{\sqrt{2}}{4}r \approx 0.354\, r.$$

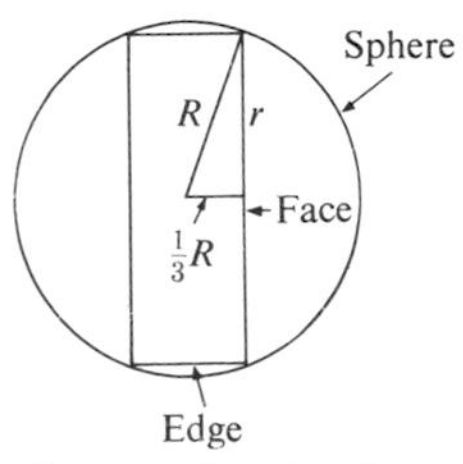

Cross section showing relation between radius R of sphere and radius r of coin.

And so the coin should be about 35% as thick as the diameter of the coin.

Digression: A Note on the Principle of Symmetry when Points Are Dropped on a Line

Suppose that several points are dropped at random on the unit interval from 0 to 1. For example, suppose w, x, y are these points as shown in the figure. These three points divide the interval into four segments with lengths x, $y - x$, $w - y$, and $1 - w$. When three points are dropped at random repeatedly, each drop of three produces four segments, and each segment (leftmost, second, third, and rightmost) has a distribution across these drops. It is easy to find the cumulative distribution of the length of the leftmost interval. Consider some number t. What is the chance that all three points fall to the right of it? Since the three points are dropped independently and each has a chance of $1 - t$ of falling to the right of t, the answer is $(1 - t)^3$. Thus

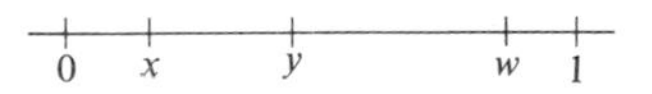

Three points dropped on the unit interval.

$$P\,(\text{leftmost point is to right of } t) = (1 - t)^3.$$

Example. What is the median position of the leftmost point? The median is the position that is exceeded half the time. We want $(1 - t)^3 = \frac{1}{2}$. The appropriate root is given by

$$1 - t = \sqrt[3]{\tfrac{1}{2}}, \qquad \text{and so} \qquad t \approx 0.206.$$

While the calculation of the distribution for the length of the leftmost segment is easy, and you could get the rightmost one by symmetry, you might boggle at finding the distribution for the second or third segment. You may have guessed already that they are the same as that for the leftmost segment, but most people do not. It is the purpose of the next remarks to make that proposition reasonable.

Instead of dropping points on a unit interval, let us drop them on a circle of unit circumference; instead of three points, let us drop four and call the fourth one z.

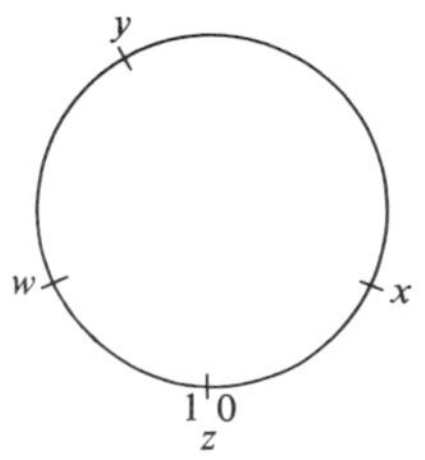

Four points dropped on a circle of unit circumference.

Thus, the points x, y, w are dropped at random on a unit interval as before, but it does not have a scale on it. We drop the fourth point z, also at random. The four points have all been treated alike, and the four segments of the circle, here (zx), (xy), (yw), and (wz), arise from a process equitable to all segments. Imagine dropping four points many times and each time getting the distance from z to the first counterclockwise point, from there to the next, and so on. Then we would generate four distributions of segment lengths, and these distributions would be alike across many drops of 4 points.

Now for each drop, cut the circle at z, and straighten it into a unit interval labeling the ends 0 and 1 as implied by the figure. The drop of four points on a circle using z as a cut is equivalent to a drop of three points on the unit interval.

Although you may still have lingering doubts, we shall not give a formal proof, but we do state the principle.

PRINCIPLE OF SYMMETRY: *When n points are dropped at random on an interval, the lengths of the $n + 1$ line segments have identical distributions.*

39. The Clumsy Chemist

In a laboratory, each of a handful of thin 9-inch glass rods had one tip marked with a blue dot and the other with a red. When the laboratory assistant tripped and dropped them onto the concrete floor, many broke into three pieces. For these, what was the average length of the fragment with the blue dot?

Solution for The Clumsy Chemist

Assuming these rods broke at random, the principle of symmetry says that each fragment—blue-dotted, middle, and red-dotted segment—would have the same distribution and the same mean. Since the means have to add to 9 inches, the blue-dotted segments average about 3 inches.

40. The First Ace

Shuffle an ordinary deck of 52 playing cards containing four aces. Then turn up cards from the top until the first ace appears. On the average, how many cards are required to produce the first ace?

Solution for The First Ace

Assume that the principle of symmetry holds for discrete as well as continuous events. The four aces divide the pack into 5 segments of size from 0 to 48 cards. If two aces are side by side, we say the segment between them is of length 0. If the first card is an ace, the segment before it is of length zero, and similarly for the segment following an ace that is a last card. The principle of symmetry says the 5 segments should average $\frac{48}{5} = 9.6$ cards. The next card is the ace itself, so it is the 10.6th card on the average.

41. The Locomotive Problem

(a) A railroad numbers its locomotives in order, 1, 2, . . . , N. One day you see a locomotive and its number is 60. Guess how many locomotives the company has.

(b) You have looked at 5 locomotives and the largest number observed is 60. Again guess how many locomotives the company has.

Discussion for The Locomotive Problem

While the questions as stated provide no "right" answers, still there are some reasonable things to do. For example, the symmetry principle discussed earlier suggests that when one point is dropped, on the average the two segments will be of equal size, and so you might estimate in part (a) that the number is 119, because the segment to the left of 60 has 59, $2(59) = 118$, and $118 + 1 = 119$.

Similarly in part (b), you might estimate that the 5 observed numbers separate the complete series into 6 pieces. Since $60 - 5 = 55$, the average length of the first 5 pieces is 11, and so you might estimate the total number as 60 + 11 or 71. Of course, you cannot expect your estimate to be exactly right very often.

The method just described makes sure that in many such estimates you average close to the correct value. That is, imagine many problems in which the unknown number N is to be guessed. Follow the estimation program described above each time (draw a sample, make an estimate). Then the set of estimates will average close to the true value in the long run.

On the other hand, you might not be interested in being close in the long run, or in averaging up well. You might want to try to be exactly right this time, however faint the hope. Then a reasonable strategy is just to guess the largest number you have seen. If you've seen 2 locomotives, then the chance that a sample of 2 contains the largest is $(N - 1)/\binom{N}{2}$ or $\frac{2}{N}\cdot$

The method of confidence limits is often used to make an interval estimate. For a description, I will confine myself to the case of one observation. If the company has N locomotives and we draw a random one, then the probabilities of the numbers $1, 2, \ldots, N$ are each $1/N$. Therefore we can be sure that the chance that our locomotive is in some set is the size of the set divided by N. For example, let n be the random number to be drawn, then for even values of N, $P(n > N/2) = \frac{1}{2}$, and for odd values of N the probability is slightly more. Then we can read the statement $n > N/2$ and say that the probability is at least $\frac{1}{2}$ that it is true when n is a random variable. If we have observed the value of n and do not know N, but wanted to say something about it, we could say $2n > N$, and that would put an upper bound on N. The statement itself is either right or wrong in any particular instance, and it is right in more than half of experiments and statements made in this manner. If one wanted to be surer, then one could change the limits. For example, $P(n \geq \frac{1}{3}N) \geq \frac{2}{3}$. The confidence statement would be $3n \geq N$, and we would be at least $\frac{2}{3}$ sure it was correct. In our problem, if we wanted to be at least $\frac{2}{3}$ sure of making a statement that contains the correct value of N, we say N is between 60 and 180.

Another method of estimation that is much in vogue is maximum likelihood. One would choose the value of N that makes our sample most likely. For example, if $N = 100$, our sample value of 60 would have probability $\frac{1}{100}$; but if $N = 60$, its probability is $\frac{1}{60}$. We can't go lower than 60 because if $N = 59$ or less we can't observe 60, and our sample would have probability 0. Consequently, if n is the observed value, the maximum likelihood estimate of N is n.

In this discussion I have not tried to use casual information such as "it's a large company, and so it must have at least 100 locomotives, but it couldn't possibly have 100,000." Such information can be useful.

42. The Little End of the Stick

(a) If a stick is broken in two at random, what is the average length of the smaller piece?

(b) (For calculus students.) What is the average ratio of the smaller length to the larger?

Solution for The Little End of the Stick

(a) Breaking "at random" means that all points of the stick are equally likely as a breaking point (uniform distribution). The breaking point is just as likely to be in the left half as the right half. If it is in the left half, the smaller piece is on the left; and its average size is half of that half, or one-fourth the length of the stick. The same sort of argument applies when the break is in the right half of the stick, and so the answer is one-fourth of the length.

(b) We might suppose that the point fell in the right-hand half. Then $(1 - x)/x$ is the fraction if the stick is of unit length. Since x is evenly distributed from $\frac{1}{2}$ to 1, the average value, instead of the intuitive $\frac{1}{3}$, is

$$2\int_{\frac{1}{2}}^{1} \frac{1-x}{x}\,dx = 2\int_{\frac{1}{2}}^{1}\left(\frac{1}{x} - 1\right)dx$$
$$= 2\log_e 2 - 1 \approx 0.386.$$

43. The Broken Bar

A bar is broken at random in two places. Find the average size of the smallest, of the middle-sized, and of the largest pieces.

Solution for The Broken Bar

We might as well work with a bar of unit length. Let x and y be the positions of the two breaking points, x the leftmost one (Fig. 1). We know from the principle of symmetry that each of the three segments (left, middle, and right) averages $\frac{1}{3}$ of the length in repeated drops of two points. But we are asked about the smallest one, for example. If we drop two points at

Fig. 1. Interval with break points x and y.

random, let X stand for the position of the first point dropped and Y for the position of the second. Then the random pair (X, Y) is uniformly distributed over a unit square as in Fig. 2, and probabilities can be measured by areas. For example, the probability that $X < 0.2$ and $Y < 0.3$ is given by the area below and to the left of $(0.2, 0.3)$, and it is $0.2 \times 0.3 = 0.06$.

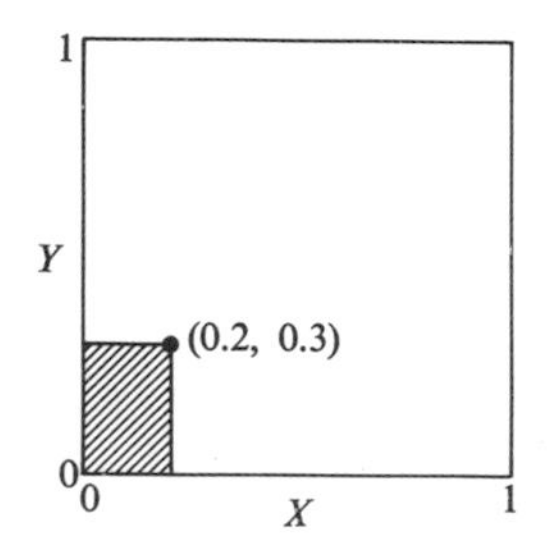

Fig. 2. Unit square representing probability distribution for a pair of points (X, Y) dropped on a unit interval.

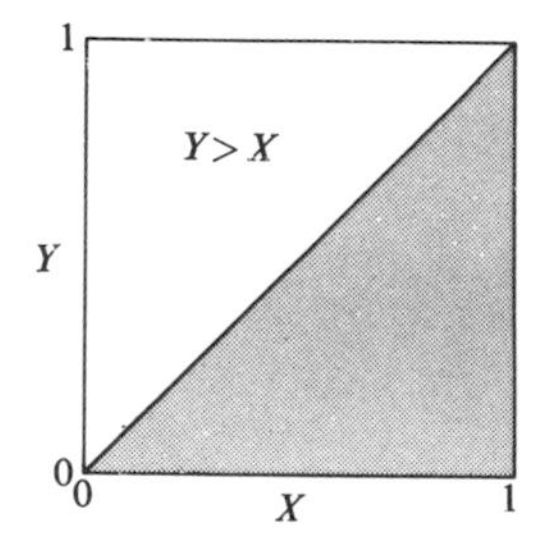

Fig. 3. Unshaded area shows where $Y > X$.

For convenience, let us suppose that X is to the left of Y, or that $X < Y$. Then the distribution is over the unshaded half-square in Fig. 3. Then probabilities are still proportional to areas, but the area must be multiplied by 2 to get the probability. If we want to get the average length for the segment of smallest length, then note that either X, $Y - X$, or $1 - Y$ is smallest. Let us suppose X is smallest, so that

$$X < Y - X \quad \text{or, equivalently,} \quad 2X < Y,$$

and

$$X < 1 - Y \quad \text{or, equivalently,} \quad X + Y < 1.$$

In Fig. 4, the triangular region meeting all these conditions is shown heavily outlined. Although X runs from 0 to $\frac{1}{3}$, it must be averaged over the triangular region. The key fact from plane geometry is that the centroid of a triangle is $\frac{1}{3}$ of the way from a base toward the opposite vertex. The base of interest in the heavily outlined triangle is the one on the Y-axis. The altitude parallel to the X-axis is $\frac{1}{3}$. Consequently, the mean of X is $\frac{1}{3} \cdot \frac{1}{3} = \frac{1}{9}$. Therefore the average value of the smallest segment is $\frac{1}{9}$.

Let's see what happens if X is the largest. We want

$$X > Y - X \quad \text{or, equivalently,} \quad 2X > Y,$$

and

$$X > 1 - Y \quad \text{or, equivalently,} \quad X + Y > 1.$$

Figure 5 shows the appropriate quadrilateral region heavily outlined. To get its mean for X, we break the quadrilateral into two triangles along the

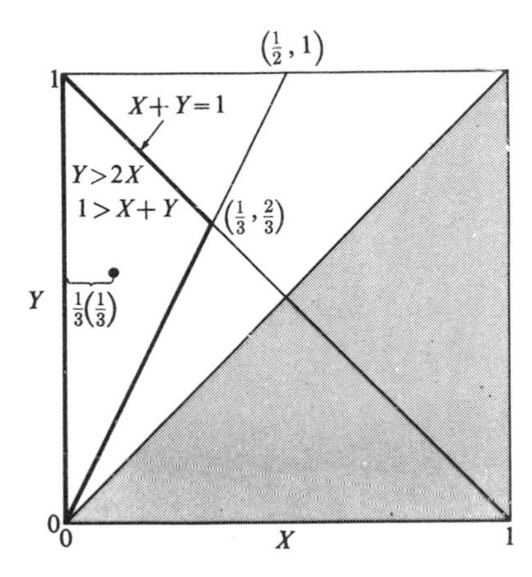

Fig. 4. Triangular region where leftmost segment is smallest is heavily outlined.

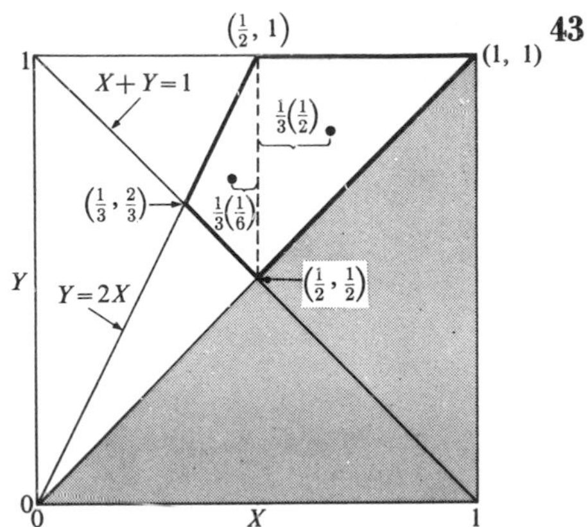

Fig. 5. Region where X is greatest is heavily outlined.

dotted line. Then we compute the mean for X for each triangle separately and weight the two means by the areas of the triangles to get the final answer.

The mean of X for the right-hand triangle whose base is the dotted line is $\frac{1}{2} + \frac{1}{3}(\frac{1}{2})$. That for the left-hand triangle whose base is the dotted line is $\frac{1}{2} - \frac{1}{3}(\frac{1}{6})$. The weights are proportional to the altitudes $\frac{1}{2}$ and $\frac{1}{6}$, respectively, because the triangles have a common base. Finally, the mean of X is

$$\frac{\frac{1}{2}(\frac{1}{2} + \frac{1}{6}) + \frac{1}{6}(\frac{1}{2} - \frac{1}{18})}{\frac{1}{2} + \frac{1}{6}} = \frac{11}{18}.$$

Since the mean of the smallest is $\frac{1}{9}$ or $\frac{2}{18}$ and that for the largest $\frac{11}{18}$, the mean for the middle segment is $1 - \frac{11}{18} - \frac{2}{18} = \frac{5}{18}$. You may want to check this by applying the method just used when, for example, $1 - Y > X > Y - X$.

So finally, the means of the smallest, middle-sized, and longest segments of the broken bar are proportional to 2, 5, and 11, respectively.

When we break a bar in 2 pieces, the average lengths of the smaller and larger pieces are proportional to

$$\frac{1}{4}, \frac{3}{4}, \quad \text{which can be written} \quad \frac{1}{2}(\frac{1}{2}), \frac{1}{2}(\frac{1}{2} + \frac{1}{1}).$$

For 3 pieces we have, in order, the proportions

$$\frac{1}{9}, \frac{5}{18}, \frac{11}{18},$$

or

$$\frac{1}{3}(\frac{1}{3}), \frac{1}{3}(\frac{1}{3} + \frac{1}{2}), \frac{1}{3}(\frac{1}{3} + \frac{1}{2} + \frac{1}{1}).$$

In general, if there are n pieces the average lengths in order of size are proportional to

$$\text{smallest:} \quad \frac{1}{n}\left(\frac{1}{n}\right)$$

$$\text{next largest:} \quad \frac{1}{n}\left(\frac{1}{n} + \frac{1}{n-1}\right)$$

$$\text{3rd:} \quad \frac{1}{n}\left(\frac{1}{n} + \frac{1}{n-1} + \frac{1}{n-2}\right)$$

$$\vdots$$

$$\text{largest:} \quad \frac{1}{n}\left(\frac{1}{n} + \frac{1}{n-1} + \cdots + \frac{1}{2} + 1\right).$$

But I have no easy proof of this.

44. Winning an Unfair Game

A game consists of a sequence of plays; on each play either you or your opponent scores a point, you with probability p (less than $\frac{1}{2}$), he with probability $1 - p$. The number of plays is to be even—2 or 4 or 6 and so on. To win the game you must get *more than* half the points. You know p, say 0.45, and you get a prize if you win. You get to choose in advance the number of plays. How many do you choose?

Solution for Winning an Unfair Game

Don't balk just because the game is unfair; after all you are the only one eligible for a prize. Let us call you player A and your opponent player B. Let the total number of plays be $N = 2n$. On a given play, your chance of winning a point is p, your opponent's $q = 1 - p$.

At first blush, most people notice that the game is unfair and therefore that, as N increases, the expected value of the difference (A's points $-$ B's points) grows more and more negative. They conclude that A should play as little as he can and still win—that is, two plays.

Had an odd number of plays been allowed, this reasoning based on expected values would have led to the correct answer, and A should choose only one play. With an even number of plays, two opposing effects are at work: (1) the bias in favor of B, and (2) the redistribution of the probability in the middle term of the binomial distribution (the probability of a tie) as the number of plays increases.

Consider, for a moment, a fair game ($p = \frac{1}{2}$). Then the larger N, the larger A's chance to win because as $2n$ increases, the probability of a tie tends to zero, and the limiting value of A's chance to win is $\frac{1}{2}$. For $N = 2, 4, 6$, his probabilities are $\frac{1}{4}, \frac{5}{16}, \frac{22}{64}$. Continuity suggests that for p slightly less than $\frac{1}{2}$, A should choose a large but finite number of plays. But if p is small, $N = 2$ should be optimum for A. It turns out that for $p < \frac{1}{3}$, $N = 2$ is optimum.

Your probability of winning in a game of $2n$ trials is the sum of the probabilities of getting $n + 1, n + 2, \ldots, 2n$ points, a sum given by

$$P_{2n} = \sum_{x=n+1}^{2n} \binom{2n}{x} p^x q^{2n-x}.$$

In a game of $2n + 2$ plays, the probability of winning at least $n + 2$ points and the game is

$$P_{2n+2} = \sum_{x=n+2}^{2n+2} \binom{2n+2}{x} p^x q^{2n+2-x}.$$

A game composed of $2n + 2$ plays can be regarded as having been created by adding two plays to a game of $2n$ plays. Unless player A has won either n or $n + 1$ times in the $2n$ game, his status as a winner or loser cannot differ in the $2n + 2$ game from that in the $2n$ game.

Except for these two possibilities, P_{2n+2} would be identical with P_{2n}. These exceptions are: (1) having $n + 1$ successes in the first $2n$ plays, A loses the next two, thus reducing his probability of winning in the $2n + 2$ game by

$$q^2 \binom{2n}{n+1} p^{n+1} q^{n-1};$$

or (2) having won n plays in the $2n$ game, he wins the next two, increasing his probability by

$$p^2 \binom{2n}{n} p^n q^n.$$

If $N = 2n$ is the optimum value, then both $P_{N-2} \leq P_N$ and $P_N \geq P_{N+2}$ must hold. The results of the previous paragraph imply that these inequalities are equivalent to the following two inequalities:

$$(1) \qquad \begin{aligned} q^2 \binom{2n-2}{n} p^n q^{n-2} &\leq p^2 \binom{2n-2}{n-1} p^{n-1} q^{n-1}, \\ q^2 \binom{2n}{n+1} p^{n+1} q^{n-1} &\geq p^2 \binom{2n}{n} p^n q^n. \end{aligned}$$

After some simplifications, which you may wish to verify (we exclude the trivial case $p = 0$), we reduce inequalities (1) to

$$(2) \qquad (n-1)q \leq np; \qquad nq \geq (n+1)p.$$

These inequalities yield, after a little algebra, the condition

$$(3) \qquad \frac{1}{1 - 2p} - 1 \leq 2n \leq \frac{1}{1 - 2p} + 1.$$

Thus unless $1/(1 - 2p)$ is an odd integer, N is uniquely determined as the nearest even integer to $1/(1 - 2p)$. When $1/(1 - 2p)$ is an odd integer, both adjacent even integers give the same optimum probability. And we can incidentally prove that when $1/(1 - 2p) = 2n + 1$, $P_{2n} = P_{2n+2}$.

Consequently for $p = 0.45$, we have $1/(1 - 0.9) = 10$ as the optimum number of plays to choose.

This material is abbreviated from "Optimal length of play for a binomial game," *Mathematics Teacher*, Vol. 54, 1961, pp. 411–412.

P. G. Fox originally alluded to a result which gives rise to this game in "A primer for chumps," which appeared in the *Saturday Evening Post*, November 21, 1959, and discussed the idea further in private correspondence arising from that article in a note entitled "A curiosity in the binomial expansion—and a lesson in logic." I am indebted to Clayton Rawson and John Scarne for alerting me to Fox's paper and to Fox for helpful correspondence.

45. Average Number of Matches

The following are two versions of the matching problem:

(a) From a shuffled deck, cards are laid out on a table one at a time, face up from left to right, and then another deck is laid out so that each of its cards is beneath a card of the first deck. What is the average number of matches of the card above and the card below in repetitions of this experiment?

(b) A typist types letters and envelopes to n different persons. The letters are randomly put into the envelopes. On the average, how many letters are put into their own envelopes?

Solution for Average Number of Matches

Let us discuss this problem for a deck of cards. Given 52 cards in a deck, each card has 1 chance in 52 of matching its paired card. With 52 opportunities for a match, the expected number of matches is $52(\frac{1}{52}) = 1$; thus, on the average you get 1 match. Had the deck consisted of n distinct cards, the expected number of matches would still be 1 because $n(1/n) = 1$. The result leans on the theorem that the mean of a sum is the sum of the means.

More formally, each pair of cards can be thought of as having associated with it a random variable X_i that takes the value 1 when there is a match and the value 0 when there is not. Then

$$E(X_i) = 1\left(\frac{1}{n}\right) + 0\left(1 - \frac{1}{n}\right) = \frac{1}{n}\cdot$$

Finally, the total number of matches is $\sum X_i$, and the expected value of a sum is the sum of the expected values, and so

$$E\left(\sum_{i=1}^{n} X_i\right) = \sum_{i=1}^{n} E(X_i) = \sum_{i=1}^{n} \frac{1}{n} = \frac{n}{n} = 1,$$

as before.

46. Probabilities of Matches

Under the conditions of the previous matching problem, what is the probability of exactly r matches?

Solution for Probabilities of Matches

This problem looks like a first cousin to the Poisson problem we discussed earlier (Problem 28). Whereas that counterfeit coin problem had *independent* chances of a bad coin (or a 1) in each position, in the current problem the chances of a match are not independent for each pair. For example, if we have $n - 1$ pairs of cards matching, the nth pair must surely match too, and so we do not have independence. Nevertheless, for large values of n, the degree of dependence is not great, and so we should anticipate that the probability of r matches in this problem may agree fairly closely with the probability of r counterfeit coins given by the Poisson, because in both cases we have many opportunities for events of small probability to occur. In the end then we want to compare the answer to the current problem with that for the Poisson with mean equal to 1.

To get started on such problems, I like to see results for small values of n, because often they are suggestive. For $n = 1$, a match is certain. For $n = 2$, the probability is $\frac{1}{2}$ of 0 matches and $\frac{1}{2}$ of 2 matches. For $n = 3$: let us number the cards 1, 2, and 3 and lay out in tabular form the 6 permutations for the matching deck beneath the target deck, which might as well be in serial order.

ARRANGEMENTS AND MATCHES, $n = 3$

Target deck:	1	2	3	Number of matches
Permutations of matching deck	1	2	3	3
	1	3	2	1
	2	1	3	1
	2	3	1	0
	3	1	2	0
	3	2	1	1

Summarizing the results of the listing gives us

DISTRIBUTION OF NUMBER OF MATCHES, $n = 3$

Number of matches	0	1	2	3
Probability	$\frac{2}{6}$	$\frac{3}{6}$	$\frac{0}{6}$	$\frac{1}{6}$

I also wrote out the 24 permutations for $n = 4$. Examine the Summary Table for $n = 1, 2, 3$, and 4. Observe that the probability of n matches is $1/n!$, because only one among the $n!$ permutations gives n matches.

SUMMARY TABLE

Number of matches:	0	1	2	3	4
$n = 1$, Probability	0	1			
$n = 2$, Probabilities	$\frac{1}{2}$	0	$\frac{1}{2}$		
$n = 3$, Probabilities	$\frac{2}{6}$	$\frac{3}{6}$	0	$\frac{1}{6}$	
$n = 4$, Probabilities	$\frac{9}{24}$	$\frac{8}{24}$	$\frac{6}{24}$	0	$\frac{1}{24}$

Note, too, that the mean for each distribution is 1, as advertised in the previous problem—this gives a partial check on the counting.

We need to break into the problem somehow—get a relation between some probability in the table and some other probability. Let $P(r|n)$ be the probability of exactly r matches in an arrangement of n things. We could get exactly r matches by having a specific r match and having the rest not match. For example, the probability that the first r match and the rest do not is

$$\frac{1}{n(n-1)\cdots(n-r+1)} P(0|n-r).$$

But there are $\binom{n}{r}$ mutually exclusive choices of r positions in which to have exactly r matches. Therefore

$$\begin{aligned} P(r|n) &= \binom{n}{r} \frac{1}{n(n-1)\cdots(n-r+1)} P(0|n-r) \\ &= \binom{n}{r} \frac{(n-r)!}{n!} P(0|n-r) \\ &= \frac{n!}{r!(n-r)!} \frac{(n-r)!}{n!} P(0|n-r), \end{aligned}$$

$$(1) \qquad P(r|n) = \frac{1}{r!} P(0|n-r), \qquad r = 0, 1, \ldots, n-1.$$

For $r = n$ we know $P(n|n) = 1/n!$, and so we can extend our notation by assigning $P(0|0) = 1$, if we wish, without doing violence to the other notation.

46

Let us check the relation (1) for $n = 4$, $r = 2$. For $n = 4$, $r = 2$, it says

$$P(2|4) = \frac{1}{2!} P(0|2).$$

The Summary Table gives

$$P(2|4) = \tfrac{6}{24}, \qquad P(0|2) = \tfrac{1}{2},$$

and $\frac{6}{24} = \frac{1}{4}$, which checks expression (1) for this example.

Since we deal with probability distributions, the sum of the probabilities over all the possible numbers of matches for a given value of n is 1, or, in symbols,

$$P(0|n) + P(1|n) + \cdots + P(n-1|n) + P(n|n) = 1.$$

We could rewrite this equation, using relation (1), as

$$(2)\quad \frac{1}{0!} P(0|n) + \frac{1}{1!} P(0|n-1) + \frac{1}{2!} P(0|n-2) + \cdots + \frac{1}{(n-1)!} P(0|1) + \frac{1}{n!} = 1.$$

Since we know $P(n|n) = 1/n!$, we could successively solve the equations for $P(0|n)$ for higher and higher values of n. For example, for $n = 1$ we get

$$P(0|1) + \frac{1}{1!} = 1$$

or

$$P(0|1) = 0.$$

For $n = 2$

$$P(0|2) + \frac{1}{1!} P(0|1) + \frac{1}{2!} = 1$$

or

$$P(0|2) = \tfrac{1}{2}.$$

For $n = 3$

$$P(0|3) + \frac{1}{1!} P(0|2) + \frac{1}{2!} P(0|1) + \frac{1}{3!} = 1$$

or

$$P(0|3) = \tfrac{1}{3}.$$

The foregoing examples show that in principle we can get $P(0|n)$ for any n, but they do not yet suggest a general formula for $P(0|n)$. Sometimes taking differences helps. Let's look at $P(0|n) - P(0|n-1)$ for the values of n up to 4 (given in the Summary Table):

$$P(0|1) - P(0|0) = 0 - 1 = -1 = -\frac{1}{1!},$$

$$P(0|2) - P(0|1) = \frac{1}{2} - 0 = +\frac{1}{2} = +\frac{1}{2!},$$

$$P(0|3) - P(0|2) = \frac{2}{6} - \frac{1}{2} = -\frac{1}{6} = -\frac{1}{3!},$$

$$P(0|4) - P(0|3) = \frac{9}{24} - \frac{2}{6} = +\frac{1}{24} = +\frac{1}{4!}.$$

These results suggest that the differences have the form $(-1)^r/r!$. That is,

$$P(0|n-r) - P(0|n-r-1) = (-1)^{n-r}/(n-r)!.$$

When we sum the differences, we get on the left-hand side

$$P(0|n-r) - P(0|0) = -\frac{1}{1!} + \frac{1}{2!} - \cdots + \frac{(-1)^{n-r}}{(n-r)!}.$$

Writing $P(0|0)$ in the form $1/0!$ and transposing it, we get, if the differences have the conjectured form,

$$(3) \quad P(0|n-r) = \frac{1}{0!} - \frac{1}{1!} + \frac{1}{2!} - \frac{1}{3!} + \cdots + (-1)^{n-r}\frac{1}{(n-r)!}.$$

All we need to do now is to check that this guess works. We get

$$(4) \quad \frac{1}{0!}\sum_{i=0}^{n}\frac{(-1)^i}{i!} + \frac{1}{1!}\sum_{i=0}^{n-1}\frac{(-1)^i}{i!} + \frac{1}{2!}\sum_{i=0}^{n-2}\frac{(-1)^i}{i!}$$

$$+ \cdots + \frac{1}{(n-1)!}\sum_{i=0}^{1}\frac{(-1)^i}{i!} + \frac{1}{n!}\sum_{i=0}^{0}\frac{(-1)^i}{i!}.$$

Courage, brother!

This looks like a mess, but it just needs a little petting and patting. The sum in expression (4) is made up of terms of the form

$$\frac{(-1)^i}{j!\,i!},$$

where the j's come from the multipliers ahead of the summation signs and the i's from the terms behind. Let us regroup the terms so that $i + j$ is a constant. For example, when $i + j = 3$ (assuming $n \geq 3$), we get only the terms

$$\frac{(-1)^3}{0!\,3!} + \frac{(-1)^2}{1!\,2!} + \frac{(-1)^1}{2!\,1!} + \frac{(-1)^0}{3!\,0!}.$$

If we multiply this by 3!, it becomes the more familiar

$$-\frac{3!}{0!\,3!}+\frac{3!}{1!\,2!}-\frac{3!}{2!\,1!}+\frac{3!}{3!\,0!},$$

which can be written in binomial coefficient notation:

$$-\binom{3}{0}+\binom{3}{1}-\binom{3}{2}+\binom{3}{3}.$$

And the latter is just an expansion of $(x + y)^3$ when $x = -1$ and $y = 1$, another way of saying that the sum is zero because $(-1 + 1)^3 = 0^3 = 0$.

This binomial-expansion trick can be applied for each $i + j = r$ for $r = 1, 2, \ldots, n$, and for each r we get a sum of zero. For $r = 0$, we get just one term $(-1)^0/(0!0!) = 1$. Consequently, we have verified that the conjectured solution (3) satisfies eq. (2).

Could other solutions satisfy eq. (2)? No. Our technique was just an easy way to check eq. (2). We could with a bit more bother have solved for $P(0|n)$, after setting in our guess for $P(0|1), \ldots, P(0|n-1)$, thus achieving a more formal induction proof.

Finally, substituting the result (3) into result (1), we can write

$$P(r|n) = \frac{1}{r!}\left(\frac{1}{0!} - \frac{1}{1!} + \frac{1}{2!} - \cdots + \frac{(-1)^{n-r}}{(n-r)!}\right).$$

When $n - r$ is large, the parentheses contain many terms in the series expansion of e^{-1}, and so

$$P(r|n) \approx \frac{1}{r!}e^{-1}, \qquad \text{for } n - r \text{ large.}$$

We foresaw initially a close relation between the probability of r matches in the matching problem and that for a count of r in the Poisson problem with mean equal to 1. For close agreement, we have found that $n - r$ needs to be large, not just n as we originally conjectured.

The probability of 0 matches then is about $e^{-1} \approx 0.368$ for large n.

47. Choosing the Largest Dowry

The king, to test a candidate for the position of wise man, offers him a chance to marry the young lady in the court with the largest dowry. The amounts of the dowries are written on slips of paper and mixed. A slip is drawn at random and the wise man must decide whether that is the largest dowry or not. If he decides it is, he gets the lady and her dowry if he is correct; otherwise he gets nothing. If he decides against the amount written on the first slip, he must choose or refuse the next slip, and so on until he chooses one or else the slips are exhausted. In all, 100 attractive young ladies participate, each with a different dowry. How should the wise man make his decision?

Solution for Choosing the Largest Dowry

The great question is whether or not his chances are much larger than $\frac{1}{100}$. Many people suggest the strategy of passing the first half of the slips and then choosing the first one that is better than all previous draws if such a one should present itself. That is quite sensible, though not best. Few have any idea of the size of the probability of winning.

Let us begin by looking at some small problems. Since we know nothing about the numbers on the slips, we might as well replace them by their ranks. If we had 3 slips, their ranks would be 1, 2, 3, and let us say that rank 3 is the largest. Had there been but 1 or 2 slips, there is no problem for he wins against one slip, and he has a 50-50 chance of winning against 2 slips. With 3 slips the 6 possible orders of drawing slips are:

1 2 3	2 3 1*
1 3 2*	3 1 2
2 1 3*	3 2 1

One strategy passes the first slip and then chooses the first slip that later exceeds it, if any do. This strategy wins in the 3 starred orders, or half the time—an improvement over a random guess, like taking the first one.

Suppose there are four slips. The 24 orders are

1 2 3 4	2 1 3 4	3 1 2 4*†	4 1 2 3
1 2 4 3†	2 1 4 3*†	3 1 4 2*†	4 1 3 2
1 3 2 4†	2 3 1 4†	3 2 1 4*†	4 2 1 3
1 3 4 2†	2 3 4 1†	3 2 4 1*†	4 2 3 1
1 4 2 3*	2 4 1 3*	3 4 1 2*	4 3 1 2
1 4 3 2*	2 4 3 1*	3 4 2 1*	4 3 2 1

We certainly should pass the first one. We could take the first higher that appears after it, if any do. Call this plan Strategy 1. The starred items on the list show when this strategy wins. Its probability of winning is $\frac{11}{24}$, a good deal larger than the $\frac{1}{4}$ a guess would give.

In Strategy 2, pass the first 2 items and take the first higher one after that. The 10 orders in which this strategy succeeds have a dagger beside them. Strategy 1 wins more often.

It we continue to study all the permutations by listing them, the future looks quite dreary, because for even 8 slips there are 40,320 orders. Furthermore, there might be good strategies that we are missing, though it seems unlikely. Perhaps mathematics can come to our aid.

I must emphasize that the wise man knows nothing about the distribution of the dowries. To make sure of it, let the king draw the slips and report to the wise man only the rank of a slip among those drawn thus far. Only a slip that has the largest dowry thus far is worth considering; call such a dowry a *candidate.*

I plan now an indifference argument to show that the form of the optimum strategy is to pass $s - 1$ slips and choose the first candidate thereafter. We would choose a candidate at draw i if its probability of winning exceeds the probability of winning with the best strategy at a later date; formally choose draw i if

$$(1) \quad P(\text{win with draw } i) > P(\text{win with best strategy from } i + 1 \text{ on}).$$

We shall show that the probability on the right decreases as i increases, that the probability on the left increases as i increases, and therefore that a draw arrives after which it is preferable to keep a candidate rather than to go on. Then we compute the probability of winning with such a strategy and finally find its optimum value.

Being young in the game loses us no strategies that are available later, because we can always pass slips until we get to the position where we want to be. Consequently, the probability on the right-hand side of the inequality must decrease or stay constant as i increases. At $i = 0$ its probability is the one we are looking for, the optimum probability, and at $i = n - 1$ its probability is $1/n$, because that is the chance of winning with the final draw.

The probability at draw i of a candidate's being the largest in the entire sample is the probability that the maximum is among the first i draws, namely i/n, which strictly increases with i from $1/n$ to 1. Somewhere along the line i/n exceeds the probability of winning achievable by going on. The form of the optimum strategy can thus be expressed by the rule: let the first $s - 1$ go by and then choose the first candidate thereafter. Let us compute the probability of winning with strategies of this form. The probability of a win is the probability of there being only one candidate from draw s through draw n. The probability that the maximum slip is at draw k is $1/n$. The probability that the maximum of the first $k - 1$ draws appears in the first $s - 1$ is $(s - 1)/(k - 1)$. The product $(s - 1)/[n(k - 1)]$ gives the probability of a win at draw k, $s \leq k \leq n$. Summing these numbers gives us the probability $\pi(s, n)$ of picking the true maximum of n by the optimum strategy as

$$(2) \quad \pi(s, n) = \frac{1}{n}\sum_{k=s}^{n} \frac{s-1}{k-1} = \frac{s-1}{n}\sum_{k=s-1}^{n-1} \frac{1}{k}$$

$$= \frac{s-1}{n}\left(\frac{1}{s-1} + \frac{1}{s} + \cdots + \frac{1}{n-1}\right), \qquad 1 < s \leq n;$$

since the first draw is always a candidate, $\pi(1, n) = 1/n$. Note for $n = 4$, $s = 2$, that $\pi(2, 4) = \frac{11}{24}$, as we got in our example.

The optimum value of s, say s^*, is the smallest s for which our initial inequality holds. That is the smallest s for which

$$(3) \qquad \frac{s}{n} > \pi(s+1, n) = \frac{s}{n}\left(\frac{1}{s} + \frac{1}{s+1} + \cdots + \frac{1}{n-1}\right),$$

or equivalently that s for which

$$(4) \quad \frac{1}{s} + \frac{1}{s+1} + \cdots + \frac{1}{n-1} < 1 < \frac{1}{s-1} + \frac{1}{s} + \frac{1}{s+1} + \cdots + \frac{1}{n-1}.$$

OPTIMUM VALUES OF s AND PROBABILITIES OF WINNING FOR THE DOWRY PROBLEM

n	s	$\pi(s, n)$	n	s	$\pi(s, n)$
1	1	1.000	10	4	0.399
2	1	0.500	20	8	0.384
3	2	0.500	50	19	0.374
4	2	0.458	100	38	0.371
5	3	0.433	∞	n/e	$1/e \approx 0.368$

The table gives optimum values of s and their probabilities for a few values of n. For $n = 100$, pass 37 and take the first candidate thereafter.

Large Values of n

For large n, we can approximate $\sum_{i=1}^{n} 1/i$ by $C + \log_e n$, where C is Euler's constant. Using this approximation in formula (2), we get, if s and n are large,

$$(5) \qquad \pi(s, n) \approx \frac{s-1}{n} \log_e \frac{n-1}{s-1} \approx \frac{s}{n} \log_e \frac{n}{s}.$$

Similarly, approximating the left- or right-hand sum in inequality (4) shows us that $\log_e(n/s) \approx 1$, and so $s \approx n/e$. Substituting these results into the final line of eq. (5) gives us the result

$$(6) \qquad \lim_{n\to\infty} \pi(s, n) = \frac{1}{e} \approx 0.368, \quad s \approx \frac{n}{e}.$$

To sum up, for large values of n, the optimum strategy passes approximately the fraction $1/e$ of the slips and chooses the first candidate thereafter, and then the probability of winning is approximately $1/e$.

Is it not remarkable in this game which at first blush offers a chance of about $1/n$ of winning that a simple strategy yields a probability of over $\frac{1}{3}$, even for enormous values of n?

And, of course, for either sex, the implications of these results for the choice of a marriage partner may repay careful study for those who are still single.

In the previous problem the wise man has no information about the distribution of the numbers. In the next he knows exactly.

48. Choosing the Largest Random Number

As a second task, the king wants the wise man to choose the largest number from among 100, by the same rules, but this time the numbers on the slips are randomly drawn from the numbers from 0 to 1 (random numbers, uniformly distributed). Now what should the wise man's strategy be?

Solution for Choosing the Largest Random Number

The very first number could be chosen if it were large enough, for example, 0.99900, because the chance of getting a number larger than this later, let alone choosing it, is only $1 - (0.999)^{99} \approx 0.1$.

As before, we have to choose between a candidate on hand and the chance that a later number will be larger *and we will choose it*. We work back from the end. If we have not chosen before the last draw, we choose it, and it wins or loses. If we have not chosen before the next-to-last draw and it is a candidate (largest so far), we choose it if it is larger than $\frac{1}{2}$, reject it if it is less, and are indifferent to $\frac{1}{2}$ itself. If it were less than $\frac{1}{2}$, we have a better chance of winning if we go on.

If we have gotten to the third draw from the end and if we have a candidate x as the value on the slip, the probabilities of 0, 1, or 2 larger numbers later are x^2, $2x(1 - x)$, and $(1 - x)^2$, respectively. If we choose the next number larger than x, the probability of winning later is

$$2x(1 - x) + \tfrac{1}{2}(1 - x)^2,$$

because if there are 0 later, we cannot win by going on; if 1, we are sure to win; and if 2 are larger than x, the chance is only $\frac{1}{2}$ that we choose the larger. If I am indifferent to a number at some draw, I would not be indifferent to it at a later draw; instead, I would want to choose it because I do not have as many opportunities to improve my holdings as I did earlier. Consequently, when two numbers larger than the indifference number x are present later in the sequence, we can be sure I would choose the first one. It has only a 50-50 chance of being the larger of the two. Thus, in computing what would happen if we decline to choose an indifference number that has been drawn, we can be sure, in general, that the best strategy chooses the next drawing whose value exceeds the indifference number in hand.

We want to determine the value of x to which we are indifferent. For the third position from the end, it is the value that satisfies

$$x^2 = 2x(1 - x) + \tfrac{1}{2}(1 - x)^2.$$

Here x^2 is the chance of winning with the number x, and the right-hand side gives the chance of winning if we pass the x in hand. The indifference number works out to be

$$x = \frac{1 + \sqrt{6}}{5} \approx 0.6899.$$

So we choose a candidate third from last if its value exceeds 0.6899.

More generally, if there are r draws to go and we have a candidate in hand, we choose the draw if it exceeds the indifference value x computed from

$$(1) \qquad x^r = \binom{r}{1} x^{r-1}(1 - x) + \frac{1}{2}\binom{r}{2} x^{r-2}(1 - x)^2 + \cdots + \frac{1}{r}\binom{r}{r}(1 - x)^r.$$

We can solve this equation numerically using binomial tables or other devices to find values of x for modest values of r. The table of indifference numbers shows some of these.

TABLE OF INDIFFERENCE NUMBERS AND THEIR APPROXIMATIONS

Number left	Solution of Eq. (1)	$\dfrac{r}{r + \alpha}$
1	0.5000	0.5542
2	0.6899	0.7132
3	0.7758	0.7886
4	0.8246	0.8326
5	0.8559	0.8614
6	0.8778	0.8818
7	0.8939	0.8969
8	0.9063	0.9086
9	0.9160	0.9180
10	0.9240	0.9256
11	0.9305	0.9319
12	0.9361	0.9372
13	0.9408	0.9417
14	0.9448	0.9457
15	0.9484	0.9491

To go at it more approximately, we might note that as r increases, $1 - x$ gets small, and a major contribution to the right-hand side of eq. (1) comes from the lead term. So

$$x^r \approx r\,x^{r-1}(1 - x), \qquad \text{or} \qquad x \approx \frac{r}{r + 1}.$$

Alternatively, we could divide eq. (1) through by x^r. Then let $z = (1 - x)/x$ to get

$$(2) \qquad 1 = \binom{r}{1} z + \frac{1}{2}\binom{r}{2} z^2 + \cdots + \frac{1}{r}\binom{r}{r} z^r.$$

Finally use eq. (2) for solutions.

Since approximately $z = 1/r$, let us set $z = \alpha(r)/r$, where $\alpha(r)$ is a function that does not change much with r. For example,

$$\begin{aligned} \alpha(1) &= 1 & \qquad \alpha(4) &= 0.8509 \\ \alpha(2) &= 0.8990 & \qquad \alpha(5) &= 0.8415 \\ \alpha(3) &= 0.8668 \end{aligned}$$

When we set $z = \alpha(r)/r$ in eq. (2) and let r grow, we have in the limit

$$(3) \qquad 1 = \alpha + \frac{\alpha^2}{2!\,2} + \frac{\alpha^3}{3!\,3} + \cdots.$$

Here α is the limiting value of $\alpha(r)$ as r grows large. We find $\alpha \approx 0.8043$. Though we could get an excellent approximation for $\alpha(r)$ now, let us settle for the limiting value and compute

$$x = \frac{r}{r + \alpha} \approx \frac{r}{r + 0.8043},$$

to get the results shown in the 3rd column of the table.

Since the present game provides more information than the one in the immediately preceding problem, the chance of winning it should be larger. If the number of slips is 2, the player chooses the first if it is greater than $\frac{1}{2}$, otherwise he goes on. His chance of winning is $\frac{3}{4}$. Increasing the number of slips from 1 to 2 has reduced his chance of winning considerably. Some geometry which I do not give shows for $n = 3$ that the probability of a win is about 0.684. For n very large, the probability of winning reduces to about 0.580.

49. Doubling Your Accuracy

An unbiased instrument for measuring distances makes random errors whose distribution has standard deviation σ. You are allowed two measurements all told to estimate the lengths of two cylindrical rods, one clearly longer than the other. Can you do better than to take one measurement on each rod? (An unbiased instrument is one that on the average gives the true measure.)

Solution for Doubling Your Accuracy

Yes. Let A be the true length of the longer one, B that for the shorter. You could lay them side by side and measure their difference in length, $A - B$, and then lay them end to end and measure the sum of their lengths, $A + B$.

Let D be your measurement of $A - B$, S of $A + B$. Then an estimate of A is $\frac{1}{2}(D + S)$, and of B is $\frac{1}{2}(S - D)$. Now $D = A - B + d$, where d is a random error, and $S = A + B + s$, where s is a random error. Consequently,

$$\tfrac{1}{2}(D + S) = \tfrac{1}{2}(A - B + A + B + d + s) = A + \tfrac{1}{2}(d + s).$$

On the average, the error $\frac{1}{2}(d + s)$ is zero because both d and s have mean zero. The variance of the estimate of A is the variance of $\frac{1}{2}(d + s)$, which is $\frac{1}{4}(\sigma_d^2 + \sigma_s^2) = \frac{1}{4}(\sigma^2 + \sigma^2) = \frac{1}{2}\sigma^2$. This value is identical with the variance for the average of a sample of two independent measurements. Thus both our measurements have contributed their full value to measuring A. In the same manner you can show that the variance of the estimate of B is also $\frac{1}{2}\sigma^2$. Consequently, taking two measurements, one on the difference and one on the sum, gives estimates whose precision is equivalent to that where 4 measurements are used, two on each rod separately.

To achieve such good results, we must be able to align the ends of the rods perfectly. If we cannot, instead of two alignments for each measurement, we have three. If each alignment contributes an independent error with standard deviation $\sigma/\sqrt{2}$, then one measurement of the sum or difference has standard deviation $\sigma\sqrt{3/2}$. Then the variance of our estimate of A would be

$$\tfrac{1}{4}[\tfrac{3}{2}\sigma^2 + \tfrac{3}{2}\sigma^2] = \tfrac{3}{4}\sigma^2 = \sigma^2/\tfrac{4}{3}.$$

Under these assumptions our precision is only as good as $1\frac{1}{3}$ independent measurements instead of 2, but still better than a single direct measurement.

We may rationalize the assignment of standard deviation $\sigma/\sqrt{2}$ to each alignment by thinking of s (or d) as composed of the sum of two independent unbiased measurement errors, each having variance $\sigma^2/2$. Then the sum of the component errors would produce the variance assumed earlier of σ^2. When we also assign the third alignment the variance $\sigma^2/2$, our model is completed.

You can read about variances of means and sums of independent variables in PWSA, pp. 318–322.

50. Random Quadratic Equations

What is the probability that the quadratic equation

$$x^2 + 2bx + c = 0$$

has real roots?

To make this question meaningful, we shall suppose that the point (b, c) is randomly chosen from a uniform distribution over a large square centered at the origin, with side $2B$ (see the figure). We solve the problem for a given value of B; then we let B grow large so that b and c can take any values.

For the quadratic to have real roots, we must have

$$b^2 - c \geq 0.$$

In the figure, we plot the parabola $b^2 = c$ and show the regions in the square, for $B = 4$, where the original equation has real roots.

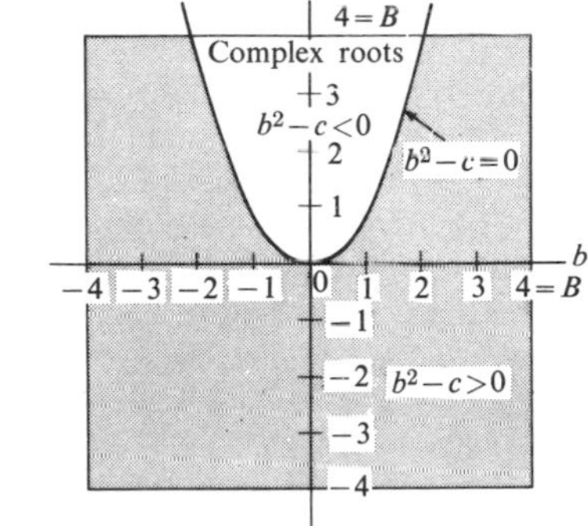

Regions yielding complex and real roots. Shaded region gives real roots; unshaded, complex.

It is an easy exercise in calculus to show that the area of the unshaded region is $\frac{4}{3}B^{\frac{3}{2}}$ (for $B \geq 1$), and, of course, the whole square has area $4B^2$. Consequently, the probability of getting complex roots is $1/(3\sqrt{B})$. When $B = 4$, the result is $\frac{1}{6}$. As B grows large, $1/\sqrt{B}$ tends to zero, and so the probability that the roots are real tends to 1!

I should warn you that the problem we have just solved is not identical with that for $ax^2 + 2bx + c = 0$. You might think you could divide through by a. You can, but if the old coefficients a, b, c were independently uniformly distributed over a cube, then b/a and c/a are neither uniformly nor independently distributed.

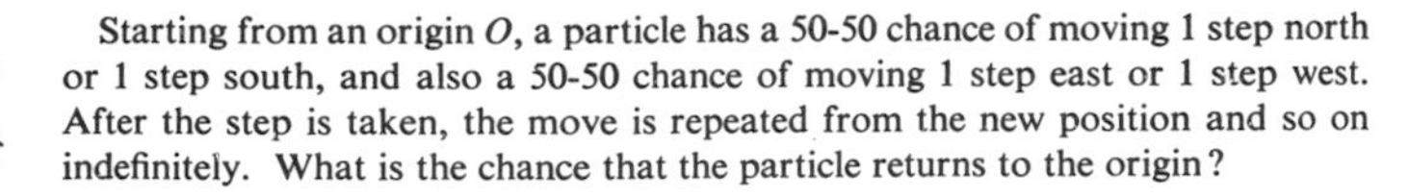

51. Two-Dimensional Random Walk

Starting from an origin O, a particle has a 50-50 chance of moving 1 step north or 1 step south, and also a 50-50 chance of moving 1 step east or 1 step west. After the step is taken, the move is repeated from the new position and so on indefinitely. What is the chance that the particle returns to the origin?

Part of lattice of points traveled by particles in the two-dimensional random walk problem. At each move the particle goes one step northeast, northwest, southeast, or southwest from its current position, the directions being equally likely.

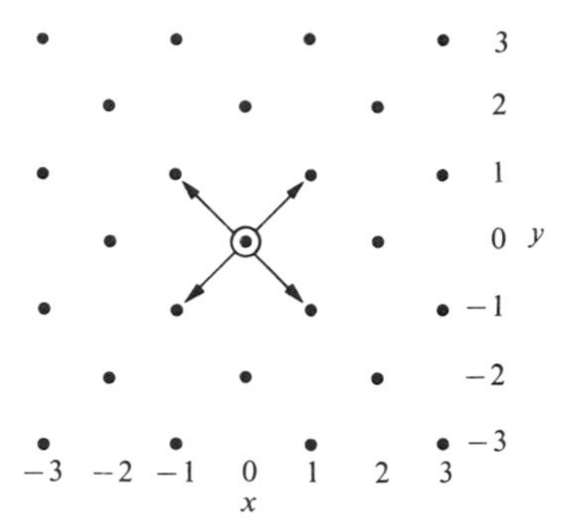

Solution for Two-Dimensional Random Walk

In the one-dimensional random walk, The Cliff-Hanger, Problem 35 (last paragraph of Solution), we found the probability that the particle returns to the origin to be unity when the probabilities of steps to the left and right were equally likely. But matters were most delicately balanced. Had either probability budged from $\frac{1}{2}$, the particle would have walked off to infinity. In two dimensions one might suppose that the particle has plenty of space to wander off to infinity. Let us see. I plan to find the average number of times a particle returns to the origin, and from this to deduce the probability that the particle returns. First, how many times will a particle come back to the origin? If P is the probability of a return, then $1 - P = Q$ is the probability of no return. The probability of exactly x returns is P^xQ, because after each return the particle might as well be regarded as starting over. If P were known, then the mean number of returns to the origin could be computed from this geometric series as

$$\mu = \sum_{x=0}^{\infty} xP^xQ.$$

Looking back at Problem 4 on trials until first success, we find the mean number to be the reciprocal of the probability of success. In that problem the success terminated the series. Here a non-return to the origin terminates the series, and so the mean number of trials to first success is $1/Q$. Consequently, the mean number of successes is $1/Q - 1$. If $Q = 1$, then the mean number of successes is 0, that is, with probability one a particle gets lost and never returns. On the other hand, the smaller Q is, the larger the mean number of returns. Indeed, for every Q there is a mean number of returns and for every mean there is a Q. If the mean number of returns before final escape were infinite (unbounded), then Q would have to vanish, and P would equal 1. More formally, as μ tends to ∞, P tends to 1. Now to get the story on the two-dimensional random walk, all we need do is compute μ.

Starting from the origin, the particle can only get home to the origin in an even number of steps. Furthermore its path can be represented as the product of two independent one-dimensional random walks, each starting at zero, one stepping east or west on each move, the other stepping north or south on each. For example, toss a coin twice; the first toss decides the east-west component, the second the north-south one. After the first two steps, the horizontal component, X, has the distribution

x	-2	0	2
$P(x)$	$\frac{1}{4}$	$\frac{2}{4}$	$\frac{1}{4}$

The vertical component, Y, is similarly distributed after two steps, and their joint probability is distributed over the 9 possible points as follows:

		$P(x, y)$			Marginal for Y y	$P(y)$
		$\cdot\ \frac{1}{16}$	$\cdot\ \frac{2}{16}$	$\cdot\ \frac{1}{16}$	2	$\frac{1}{4}$
		$\cdot\ \frac{2}{16}$	$\cdot\ \frac{4}{16}$	$\cdot\ \frac{2}{16}$	0	$\frac{2}{4}$
		$\cdot\ \frac{1}{16}$	$\cdot\ \frac{2}{16}$	$\cdot\ \frac{1}{16}$	-2	$\frac{1}{4}$
Marginal for X	x	-2	0	2		
	$P(x)$	$\frac{1}{4}$	$\frac{2}{4}$	$\frac{1}{4}$		

Joint distribution of X and Y after two moves.

The main information we wish to note is that the probability at the origin is $\frac{4}{16}$ and that it can be obtained by multiplying $P(X = 0)$ by $P(Y = 0)$ because of the independence of the component walks. Finally, we want to interpret this information. At the end of two moves $\frac{4}{16}$ of particles have returned to the origin. The contribution to the mean number of returns to the origin is then $(\frac{4}{16})1 + (\frac{12}{16})0 = \frac{4}{16}$. We compute the probability of the particle's being at the origin after 2, 4, 6, . . . trials and add these up to get the expected number of times the particle returns to the origin.

After $2n$ moves, $n = 1, 2, \ldots,$ the probability of the particle's being at the origin is

$$P(\text{particle at origin}) = P(X = 0)P(Y = 0) = \left[\binom{2n}{n}\left(\frac{1}{2}\right)^{2n}\right]^2,$$

because we must get equal numbers of east and west moves as well as equal numbers of north and south moves. (I ought really to put subscripts on X and Y, writing X_{2n}, but it makes the page look horrible to me and frightening to some.) We plan to sum, approximately, these probabilities to get the expected number of returns. For large values of n we can apply Stirling's approximation given in Problem 18 and get

$$\binom{2n}{n}\left(\frac{1}{2}\right)^{2n} = \frac{(2n)!}{n!\,n!}\left(\frac{1}{2}\right)^{2n} \approx \frac{\sqrt{2\pi}(2n)^{2n+\frac{1}{2}}e^{-2n}}{(\sqrt{2\pi}\,n^{n+\frac{1}{2}}e^{-n})^2 2^{2n}}$$
$$\approx 1/\sqrt{\pi n}.$$

For good-sized n then

$$P(\text{particle at origin}) \approx \frac{1}{\pi n}\cdot$$

We need to sum over the values of n. Recall from Problem 14 that $\sum_{n=1}^{N} 1/n \approx \log_e N$ is a number which is unbounded as N grows. What we have computed is the probability that the particle is at the origin at the end of steps numbered 2, 4, 6, 8, . . . , $2n$. Each of these probabilities is also the mean number of times the particle is at the origin at the end of exactly $2n$ trials. To get the mean total number of times the particle is at the origin, we sum because the mean of the sum is the sum of the means. Therefore the mean number of returns to the origin is unbounded, and therefore the probability of return to the origin is $P = 1$. And so each particle not only returns, but returns infinitely often. More carefully, I should say nearly every particle returns infinitely often, because there are paths such as the steady northeast course forever that allow some particles to drift off to infinity. But the fraction of such particles among all of them is zero.

52. Three-Dimensional Random Walk

As in the two-dimensional walk, a particle starts at an origin O in three-space. Think of the origin as centered in a cube 2 units on a side. One move in this walk sends the particle with equal likelihood to one of the *eight corners* of the cube. Thus, at every move the particle has a 50-50 chance of moving one unit up or down, one unit east or west, and one unit north or south. If the walk continues forever, find the fraction of particles that return to the origin.

Solution for Three-Dimensional Random Walk

Now that we know that in both one and two dimensions the particle returns to the origin with probability one, isn't it reasonable that it will surely return for any finite number of dimensions? It was to me, but I was fooled.

We have three coordinates and the probability that all three vanish at trial $2n$ is **52**

$$P(\text{particle at origin}) = P(X = 0)P(Y = 0)P(Z = 0) = \left[\binom{2n}{n}\left(\frac{1}{2}\right)^{2n}\right]^3.$$

Let's try Stirling's approximation again. We have for three dimensions after $2n$ moves

$$P(\text{particle at origin}) \approx 1/(\pi n)^{\frac{3}{2}}.$$

We can show by integration methods that $\sum 1/n^{\frac{3}{2}}$ is bounded. Replace the number $1/n^{\frac{3}{2}}$ by the area of a rectangle whose base runs from n to $n + 1$, and whose height is $1/n^{\frac{3}{2}}$. See figure. Run a smooth curve $f(n) = 1/(n - 1)^{\frac{3}{2}}$ through the upper right hand corners of the rectangles. The area under the curve exceeds the area of the rectangle:

$$\int_n^N \frac{dx}{(x-1)^{\frac{3}{2}}} = \left.\frac{-2}{(x-1)^{\frac{1}{2}}}\right]_n^N$$

$$= \frac{2}{(n-1)^{\frac{1}{2}}} - \frac{2}{(N-1)^{\frac{1}{2}}}.$$

Now as N tends to infinity, the area tends to $2/(n - 1)^{\frac{1}{2}}$, a *finite* number. This shows that the mean converges to a finite number.

We can evaluate that number by actually evaluating the early terms of the series

$$\sum_{n=1}^{\infty}\left[\binom{2n}{n}\left(\frac{1}{2}\right)^{2n}\right]^3,$$

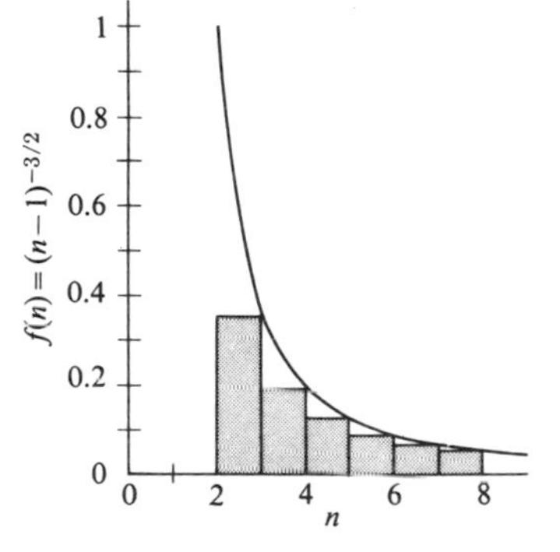

Plan for getting upper bound to $\sum_{n=k}^{\infty} 1/n^{\frac{3}{2}}$.

and then approximating the rest of the sum by integration methods. I get 0.315. After, say, 10 or 20 terms, Stirling's approximation should be very accurate, and the remainder that needs evaluation by integration is tiny by then. I used 18 terms. This 0.315 is the mean number of returns to the origin per particle. Consequently $1/Q = 1 + 0.315$, and we get

$$Q \approx 1/1.315 \approx 0.761.$$

Therefore the probability P that a particle returns to the origin is about 0.239.

For those of you who have seen the results for the random walk where the steps are to the centers of the faces of the surrounding cube rather than to the corners, you may know that the fraction returning is about 0.35.* Apparently then 8 equally likely moves reduces the chance of returning more than 6.

The same techniques for a 4-dimensional random walk where 4 coins are tossed to find the vector to be added to the present coordinates show that the probability of return is reduced to 0.105.

53. Buffon's Needle

A table of infinite expanse has inscribed on it a set of parallel lines spaced $2a$ units apart. A needle of length $2l$ (smaller than $2a$) is twirled and tossed on the table. What is the probability that when it comes to rest it crosses a line?

Solution for Buffon's Needle

This is a great favorite among geometric probability problems. The figure shows how the needle might land so that it just touches one of the parallels. We only need to look at one half-segment, because the symmetries handle the rest.

The vertical position of the needle does not matter because moving it up or down leaves the state of crossing or not crossing a vertical line unchanged. What matters is the needle's angle with the horizontal and the distance of the center of the needle from its nearest parallel. The center P is equally likely to fall anywhere between the parallels (assumption of uniform distribution); and for a fixed value of the angle θ, the chance that the line crosses one of the parallels is $2x/2a$, because the line crosses a parallel if the center falls within x units of either parallel—see the dashed needles in the figure. Because of the twirling, the angle θ might as well be thought of as uniformly distributed from 0 to $\pi/2$ radians (or in degrees from $0°$ to $90°$), because crossings that happen for angle θ also happen for angle $\pi - \theta$ (or in degrees $180° - \theta$). All we need then is the mean value of x/a, or, since $x = l \cos \theta$, the mean value of $(l/a)\cos \theta$. This average can be found by integrating

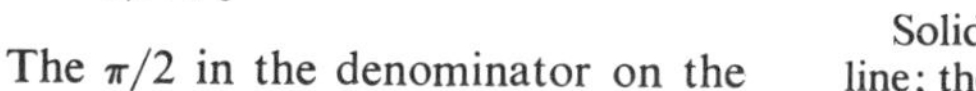

$$\frac{l/a}{\pi/2} \int_0^{\pi/2} \cos \theta \, d\theta = 2l/\pi a.$$

The $\pi/2$ in the denominator on the

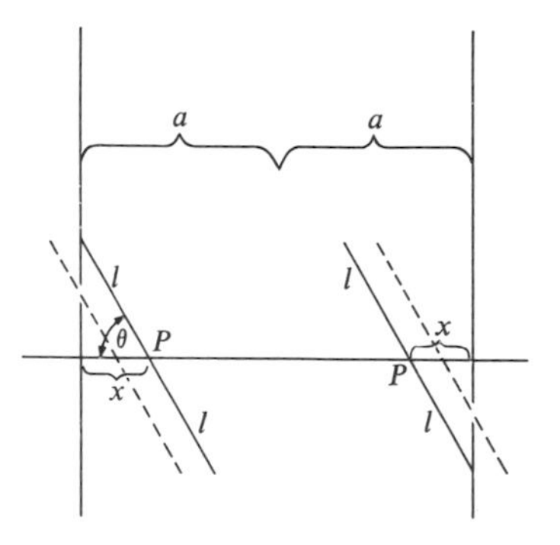

Solid needles touch one parallel line; the dashed needles cross one.

*W. Feller, *Probability theory and its applications*, 1st ed., Wiley, 1950, p. 297.

left makes the probability 1 that θ is between 0 and $\pi/2$. We can also write, remembering $2l$ is the length of the needle,

$$P(\text{needle crosses a parallel}) = \frac{2(\text{length of needle})}{\text{circumference of circle of radius } a}.$$

Why is this problem a favorite? I think because it suggests a relation between a pure chance experiment and a famous number π. You could actually construct such a table—it needn't be quite infinite—with rulings, say, an inch apart; often ordinary graph paper is so ruled. Get a needlelike object, perhaps an inch long, and keep track of the fraction of times the needle crosses a line. Then π may be estimated as about 2/(proportion of crosses). You won't get close to π very fast this way, and this estimate is always a rational number (if you get some crosses), but the charming thing is that there is any relation at all between a universal constant like π and a chance experiment. Instead of doing this experiment, wait a bit, we'll have a better one in Problem 55. If geometrical probabilities interest you, take a look at: Joseph Edwards, *Treatise on integral calculus*, Vol. II, Chelsea Publishing Co., New York, 1954 (originally printed by Macmillan in 1922). M. G. Kendall and P. A. P. Moran, *Geometrical probability*, Griffin's Statistical Monographs and Courses No. 10, Hafner Publishing Company, New York, 1963.

54. Buffon's Needle with Horizontal and Vertical Rulings

Suppose we toss a needle of length $2l$ (less than 1) on a grid with both horizontal and vertical rulings spaced one unit apart. What is the mean number of lines the needle crosses? (I have dropped $2a$ for the spacing because we might as well think of the length of the needle as measured in units of spacing.)

Solution for Buffon's Needle with Horizontal and Vertical Rulings

The mean number of vertical rulings crossed is the same as the probability of crossing a vertical ruling. From the previous problem (with $a = \frac{1}{2}$), it is $4l/\pi$. The mean number of horizontal rulings crossed must also be $4l/\pi$ because it is the same problem if you turn your head through 90°. The mean of a sum is the sum of the means, and so the mean total number of crossings is $8l/\pi$.

If the needle is of length 1, the mean number of crosses is $4/\pi \approx 1.27$.

Up to now we have worked with needles shorter than the spacing, what about longer needles?

55. Long Needles

In the previous problem let the needle be of arbitrary length, then what is the mean number of crosses?

Solution for Long Needles

Let the needle be divided into n pieces of equal lengths so that all are less than 1. If we toss each of these little needles at random, each will have a mean number of crosses obtained from the previous problem. The mean of the sum is the sum of the means, and so their expected number of crosses is 4(original length)/π. The fact that the needle was not tossed as a rigid structure does not matter to the mean.

For purposes of estimating π the experiment of tossing a long needle on a grid of squares represents a substantial improvement over the original Buffon problem. Why not get some graph paper and try it? I used a toothpick and graph paper ruled in half-inch squares. The toothpick was 5.2 half-inches long. I decided on 10 tosses, got 8, 6, 7, 6, 5, 6, 7, 5, 5, 7 crosses, totaling 62. My estimate for π is $4(5.2)/(62/10) \approx 3.35$, instead of 3.14. A friend of mine also made 10 tries, producing 67 crosses, yielding the estimate 3.10.

56. Molina's Urns

Two urns contain the same total numbers of balls, some blacks and some whites in each. From each urn are drawn n (≥ 3) balls with replacement. Find the number of drawings and the composition of the two urns so that the probability that all white balls are drawn from the first urn is equal to the probability that the drawing from the second is either all whites or all blacks.

Discussion for Molina's Urns

E. C. Molina invented this problem to display Fermat's famous conjecture in number theory as a probability problem.

Let z be the number of white balls in the first urn, x the number of whites and y the number of blacks in the second. Then we want to find integers n, x, y, and z so that

$$\left(\frac{z}{x+y}\right)^n = \left(\frac{x}{x+y}\right)^n + \left(\frac{y}{x+y}\right)^n,$$

or

$$z^n = x^n + y^n.$$

Although, for many values of n, it is known that this equation cannot be satisfied, it is not known whether it is impossible for all values of $n \geq 3$. But it is known to be impossible for $n < 2000$.